The SYSTEMS THINKING PLAYBOOK

EXERCISES TO STRETCH AND BUILD

BUILD

LEARNING AND SYSTEMS THINKING CAPABILITIES

by
Linda Booth Sweeney
& Dennis Meadows

CHELSEA GREEN PUBLISHING
WHITE RIVER JUNCTION, VERMONT

In Appreciation

The Systems Thinking Playbook was inspired and created by many people. We'd like to extend our thanks for the ideas, criticism, technical expertise, and nurturing that came generously from the following:

Prinny Anderson, Michelle Boos, Andy Bryner, Michael Buehler, Roslie Capper
Myrna Caselbolt, Sarita Chawla, Kristin Cobble, Diane Corey, Sheryl Erickson
Janet Gould, Chérie Hafford, Robert Hanig, Jennifer Hirsch, Robert Kegan
Daniel Kim, Art Kleiner, Mara Levin, Nancy Margulies, Dawna Markova
Pam McPhee, Kellie Wardman O'Reilly, Diane Reed, George Roth
Stephanie Ryan, Michael Seever, Terri Seever, Amy Seif
Don Seville, John Shibley, Cindi Stannard

And to the "thought leaders" who provided inspiration and information
for much of **The Systems Thinking Playbook:**

Chris Argyris, Mary Catherine Bateson, Gregory Bateson
Fritjof Capra, Victor Freidman, Jay Forrester
Robert Fritz, Rick Karash, Fred Kofman
Joanna Macy, George Roth, Barry Richmond
Edgar Schein, Donald Schön, Peter Senge
John Sterman, Margaret Wheatley, Danah Zohar

And for loving support from the "home teams":

Bob and Rosemarie Booth
John Sweeney

Suzanne MacDonald

Originally Published by Turning Point
First Chelsea Green printing 2010
Printed in the United States of America
Previous edition ISBN: 0-9666127-7-9
Chelsea Green edition ISBN: 978-1-60358-258-2

Chelsea Green Publishing Company
85 North Main Street, Suite 120
White River Junction, VT 05001
(802) 295-6300
www.chelseagreen.com

Contents

Guiding Ideas

Systems thinking is a broad term used to represent a set of methods and tools that focus on systems — rather than parts — as the context for defining and solving complex problems, and for fostering more effective learning and design. At its best, the practice of systems thinking helps us to stop operating from crisis to crisis, and to think in a less fragmented, more integrated way.

The games in this book highlight many of the concepts and ways or "habits of mind" associated with systems thinking, providing insights into these complex ideas in novel ways.

As our society enters the 21st century, we face an important educational challenge: How can we help people become engaged at all levels in learning how to think and behave in increasingly complex systems? More and more, practitioners and academics alike adhere to a simple premise when designing learning experiences: engage the mind and the body. In their powerful book, *An Unused Intelligence*, Andy Bryner and Dawna Markova warn that the Western culture of education leaves the problem-solving potential of our bodies virtually untapped. With this we wholeheartedly concur, and would add that the systems thinking and systems sensing potential of our bodies has been untapped as well.

What you experience using these exercises will depend on the skillful integration of key concepts, theory, techniques and experiential exercises; your familiarity with systems thinking concepts; and the insight and energy of the facilitator. Experientially, you will raise awareness of the "habits of mind" found in a systems thinker. And we invite you to have some serious fun while you're at it.

We have tried to set up the **Playbook** so that anyone — managers, CEOs, teachers, and professors, can read it and use it and find something meaningful. You don't need to be an organization development professional or trainer to use these exercises. In fact, we envision that, with a bit of preparation, a team will be able to open the **Playbook** and work through the exercises as they might the exercises in Senge et al.'s *The Fifth Discipline Fieldbook* (1994).

As you use these exercises, write down your experiences. What worked? What got the "ah ha!"? What did not? Were there cross-cultural issues that needed to be considered?

The Ways Of A Systems Thinker

Our experience studying and teaching systems thinking has led to the definition of a systems thinker as someone who:

- Sees the whole picture.
- Changes perspectives to see new leverage points in complex systems.
- Looks for interdependencies.
- Considers how mental models create our futures.
- Pays attention to and gives voice to the long-term.
- "Goes wide" (uses peripheral vision) to see complex cause and effect relationships.
- Finds where unanticipated consequences emerge.
- Focuses on structure, not on blame.
- Holds the tension of paradox and controversy without trying to resolve it quickly.
- Makes systems visible through causal maps and computer models
- Seeks out stocks or accumulations and the time delays and inertia they can create.
- Watches for "win/lose" mindsets, knowing they usually makes matters worse in situations of high interdependence.
- Sees oneself as part of, not outside of, the system.

The exercises are meant to promote a greater awareness of these ways of thinking, seeing and interacting with the world. They are best used within an inter-related and reinforcing design which covers theory, concepts, and models, and includes a relevant and detailed debriefing of the participants' experience.

The Role of Games in Teaching Systems Thinking

Whether you played them in your backyard when you were a kid, or on the front stoop, or in a gym, you probably have pleasant memories of fun times involving many different games — Checkers, Hide-and-Seek, Tag, and other initiatives that you and your friends made up. Because games are enjoyable, many people suspect that game-based training isn't serious. After all, isn't learning supposed to involve earnestness, hard work, and serious expressions?

Maybe not. According to *Psychology Magazine* (July/August, 1998), the playfulness inherent in games "makes them psychologically truer even than everyday life. Games solve major crises, train war heroes, and civilize us all. What the world needs is not less time for playing games, but more." Games permit us to learn about complex systems while we are interacting with others. They offer the chance to make mistakes without great consequence. And they are fun.

Games can:

- Reveal an individual's or group's unconscious way of interacting and solving problems.

- Illustrate the power of habits, paradigms and values in identifying problems, gathering data, and making decisions.

- Replicate the structure and behavior of reoccurring patterns of behavior — aka, systems archetypes.

- Offer a shared experience of a behavior or problem that can then form the focus of further modeling exercises.

- Help create a non-threatening environment in which participants test theories of effective social behavior and evaluate real decision options. In a game it is possible to make a big mistake, but walk away without enduring consequences.

- Engage participants who have a wide range of learning styles.

These special features of games have become even more important over the last few years, as the nature of teaching has changed. In the past, educators could spend two years giving students a masterful command of system dynamics. Now many people will devote only a few days to formal study of system dynamics and systems thinking concepts. Clearly, the concepts and skills that can be conveyed in this drastically reduced time period are very different from those offered through formal high school or college-level classes. But games, appropriately selected and sequenced, can still let us accomplish a great deal of learning.

Games can facilitate learning in two ways: through discovery or by confirmation. With discovery, players are given the rules of the game and then, typically, are surprised by behaviors that emerge during the play. Under these conditions, players can learn a lot by making mistakes, and sometimes, failing. For example, in the **Community Maze** exercise players discover the correct path only through making mistakes.

Under these conditions, players typically make many mistakes; often they fail in achieving their goals. Their errors also become the source of valuable insights about teamwork and communication. In **Space for Living** participants often mistakenly persist in a behavior that formerly was successful, but subsequently fails to satisfy their goals. After conditions have changed, participants' discovery of their mistakes, and their efforts to develop a new strategy are the essential learning opportunities of the game.

When a game is played in confirmation mode, players first learn new behaviors, skills, and knowledge. Then they play the game as an opportunity to practice their new understanding and demonstrate, or confirm, its effectiveness. Under these conditions, players typically make few errors and achieve great success. For instance, **Postcard Stories** and **Monologue/Dialogue** can both be used in confirmation mode to demonstrate that basic skills have been mastered.

Some relatively complex games may be played twice—first for the purpose of discovery and then, after discussion and learning, to confirm players' new knowledge. However, most of the games described in this volume are elementary and not suited to this double use.

Design Consideration

The power of these exercises can be either increased or diminished by the amount of thought put into the structure of the program design. The key questions are: How do the exercises best support the concepts you are trying to convey? Where do they best fit in? Will it make more sense to explain a concept (such as delays in systems) and then have participants physically experience a set of delays? What exercise will best build on and further the insights gained in the previous exercise? How do they meet the needs of the participants' various learning styles?

By asking these questions, we have found we are more able to create a seamless experience for participants, where theory and practice reinforce each other.

Creating an Environment for Learning

Creating a safe environment in which participants can explore their own behaviors is critical. The following are not hard and fast rules but rather salient factors we have found help create a safe environment conducive to learning.

Before The Session:

- Consider group size.
  Eight to twelve is an ideal size for all members of the group to be heard, to participate, and to produce useful group dynamics.

- Share the intent of your work at the beginning.
  Share the underpinnings of your design: will you be combining experiential and didactic approaches to reach multiple learning styles?

- Be wary of videotaping.
  At the very least, ask for the group's permission and explain how the tape will be used. Videotaping can cause participants to become very self-conscious, changing their normal behaviors and interfering with their learning.

- Pay attention to the seating arrangement.
  Circles and half moons tend to raise the level of engagement.

- Provide clear, up-front communications about the session.
  What can people expect? Who will be there? Should they wear comfortable clothes and flat shoes?

- Consider the implications of diverse backgrounds.
  You may find that some of the exercises do not translate directly into different cultures. Try to do a test run with someone from the environment in which you will be working.

During the Session:

- Encourage "whole speak" (mind, heart and spirit).
  Ask participants to slow down the pace of conversation and to speak authentically, from their heart and their head.

- Use a check-in.
  Give people a chance to introduce themselves (if appropriate) and

become more present by acknowledging "where their heads are" at that moment. A good question to ask: "What do you need to take care of or let go of to be fully present?"

- Provide participation options.
No one is required to participate or speak in a debriefing session. We like to use the phrase "challenge by choice" to remind people to participate at their own comfort level. Silence or passing should be mentioned as an acceptable option. No one should feel pressed to talk or disclose more than they feel is appropriate.

Degree of Physical Challenge

Unless otherwise noted, these exercises do not require special physical ability. In fact, most require little or no physical strength but rather the spirit of a "beginner's mind" and a willingness to participate. When asking for enthusiastic participation from the group, we remind them of the line on the old Coca-Cola bottles: "no deposit, no return."

Framing Techniques — Ways to introduce Learning Exercises

The way you "frame" an exercise can significantly influence the mindset participants bring to it and the lessons they take away. A "frame," is the story you tell or the metaphor you use to give meaning to an initiative. It includes the precise formulation of the goals, guidelines, and criteria for success. Selecting and presenting an appropriate and compelling frame is an art facilitators develop with practice.

Stephen Bacon, who wrote *The Conscious Use of Metaphor*, asserts: "The artistry lies in delivering the suggestion convincingly enough to ensure that the participants are invested in the challenge."

Your challenge as facilitator is knowing when to describe an exercise as something that corresponds to a specific issue in the home organization of a particular group, and when to go the more imaginary route and talk about "visiting space ships" or "avoiding a swamp of poison peanut butter." Knowing your group will help you decide what kind of frame to use.

There are at least three ways to frame the experiential exercises described in the *Playbook*: isomorphic, universal, and fantastical. Their differences are illustrated here using the **Moon Ball** exercise as an example. Take a moment to familiarize yourself with that exercise before reading about the following three types of frames.

Isomorphic

This kind of frame replicates the same or similar characteristics of the participants' organization. For example, if the group is comprised of bank managers, you could introduce **Moon Ball** this way: "Your group writes mortgages. In this next exercise, each successful hit of the beach ball is one successfully completed mortgage loan. Every person in the circle is in charge of a different department in the bank, and the mortgage applications must pass through each department. That is, they must be touched by each person. Your goal is to complete as many loans as you can in two minutes."

Universal

This kind of frame references an everyday life situation or event that could apply to anyone in the group, but is not necessarily group or organization-specific. "As team members, you all have a variety of skills and they are all required to solve problems. Your goal is to solve as many problems as you can in two minutes. A problem is 'solved' once each team member has made a contribution."

Fantastical

This kind of frame involves something completely out of the ordinary. You have fun taking the group out of their day-to-day experience. For example, "Your task is to shoot down as many space ships in Darth Vader's fleet as you can in two minutes. The fleet is represented by the beach ball. You shoot down one space ship when you manage to pass the ball (space ship) around the group with each member hitting it once and only once."

 ## Going for the Insights — Debriefing Tips

Debriefing is a process of guided discussion and reflection immediately following a group's exercise experience. There are many processes you and your group can follow to debrief an exercise systemically. We've found the following four-step process to be an effective procedure for debriefing many of the exercises in **The Systems Thinking Playbook.**

This four-step process serves a three-fold purpose:

- First, it organizes the debrief process into clear and simple steps. For those who are not full-time facilitators, this process can act as a simple organizing structure that increases one's comfort with the debrief and one's opportunity for a successful debrief.

- Second, this process helps the learner to develop a methodical and thorough approach to using the tools and concepts of systems thinking. Typically, a learner who has experienced several exercises accompanied by the four-step debrief process can recall the four steps without help. Ideally, the learner who approaches the next challenge systemically will remember and use the four-step process.

- Third, the process gives participants an opportunity to become familiar with such systems thinking terms as "behavior over time graphs," "causal loop diagrams," and "systems archetypes"—thus improving their fluency in the language of systems thinking.

Step I: Tell the Story

After an exercise, ask the group to "tell the story." What happened? What did they see? What did they feel? What did they experience? Record some of the key points from their comments on a flip chart or overhead. These are many of the variables from which the group will eventually create a "causal loop" diagram. This diagram captures, in the form of a closed loop, the cause and effect linkages between the variables in a system.

After **Moon Ball**, asking the question "What happened?" often elicits such responses as: "We didn't have a plan at first, but we figured it out with a few tries;" "We wanted to get better each time;" "We weren't listening to each other;" "We didn't take into account the differing physical abilities of our members."

These phrases can then be distilled to their essential meaning or "scrubbed" and turned into variables appropriate for creating the reference mode and the causal loop diagram. In **Moon Ball** some of the important phrases might be:

- skill of approach
- pressure to improve
- team learning
- time to change places

Step 2: Graph the Variables

Depict the behavior of selected variables over time (such as team learning) with a graph (known as a "Behavior Over Time Graph" or "Reference Mode Diagram"). This is an important step toward explaining and understanding the dynamics of the system experienced by a group. Considering the appropriate time horizon, ask the group to identify and draw out the reference mode diagram (in these exercises, time horizon will be represented by number of minutes or number of trials).

Step 3: Make the System Visible — Draw the Causal Loop Diagram

In a causal loop diagram, we connect the cause and effect relationships between the selected variables. A causal loop diagram is essential, as it helps to answer the question: "What structure could be causing the behavior we've depicted in the Reference Mode Diagram?" For a primer on how to draw causal loop diagrams see, "Guidelines for Drawing Causal Loop Diagrams," by Daniel Kim, in *The Systems Thinker*, Volume 3, Number 1. (www.thesystemsthinker.com/tstgdlines.html)

Step 4: Identify the Lessons

What are the insights the group has gained from the exercise? What structures (or in real life, what policies) would the group change to improve results? Where is the area of highest leverage? In the case of **Moon Ball** one structural change could be to substitute "External standards" for "Actual performance" as the most important cause of "Pressure to improve."

Selecting Exercises

We advise using the games in this book as short interventions to jump-start a learning experience or punctuate key insights within a long lecture. Stringing several exercises together will not constitute a coherent experience for participants. Rather, we suggest that you interweave thoughtful lectures, videos, case discussions, and small group conversations together with selected *Playbook* exercises.

In addition to the concepts you wish to explore, your choice of exercise will often be dictated by the conditions of play—number of participants, length of time available, attributes of the workshop space, availability of special equipment. We provide a summary matrix with descriptions of key attributes of the 30 exercises as well as the disciplines illustrated by each game, based on Peter Senge's *The Fifth Discipline:* team learning, personal mastery, shared vision, mental models and systems thinking. This matrix can be used to match our games to your conditions of play. But you may also select an appropriate initiative by considering the way each exercise illustrates principles of system change.

Donella Meadows has identified various "levers" through which participants in a system can change the behavior of the system. The following is a partial list of her intervention points followed by one or more **Playbook** exercises that illustrate that approach.

Change the rules: **Warped Juggle, Five Easy Pieces**

Add reinforcing or balancing processes: **Group Juggle, Living Loops**

Alter information flows; **Monologue/Dialogue, Community Maze**

Select a different time horizon: **Frames**

Change a paradigm or perspective: **Hands Down, Thumb Wrestling**

Enhance capacity for learning: **Moon Ball, Dog Biscuits & See Saws, Touch Base**

Improve team dialogue: **Squaring the Circle**

Alter the length of delays: **Balancing Tubes**

To take an example from the **Playbook**, **Warped Juggle** provides a valuable practice field for exploring the first lever, changing the rules (or in real life, changing policy or incentives.) By revisiting the assumed "rules" of the game, participants in the exercise find that they can make remarkable improvement, by a factor of ten or more, in achieving their shared goal.

Resources

A thoughtful discussion of different approaches to framing, along with many examples, is provided in *Book of Metaphors, Volume II,* by Michael A. Gass. This book is published by Kendall/Hunt Publishing Co. and distributed by Project Adventure, P.O. Box 100, Hamilton, MA 01936. Fax: (508) 468-7605

For more on systems thinking facilitation, see "Coaching and Facilitating of Systems Thinking" by Rick Karash (*The Systems Thinker* June/July 1996), and "Six Steps to Thinking Systemically" (*The Systems Thinker* March 1995). Both are available from Pegasus Communications Inc. (www.pegasuscom.com/newsletters.html).

Contact

We hope you enjoy this **Playbook** and find that it enhances the power of your training and educational programs. We look forward to hearing about your experiences!

Linda Booth Sweeney
Concord, Massachusetts

Dennis Meadows
Durham, New Hampshire

Summary Matrix of Games

Game	pg	# of people	equipment needed	time (min.)	mental models	team learning	systems thinking	shared vision	personal mastery
1-2-3-Go!	120	3 +	None	5-10	✓	✓			
Arms Crossed	22	None	None	2	✓	✓			✓
Avalanche	215	teams of 8-15	One long slender pole with two washers per team, tape	10-50	✓	✓	✓		
Balancing Tubes	103		One 3' paper tube or stick per person	5-15			✓		
Belief Release	96		One copy of the script	10-15				✓	✓
Chevreul's Pendulum	34		Large washer, 12" string, and paper target (one set for each participant)	30	✓			✓	✓
Circles in the Air	25		Pen for each participant	2-10	✓		✓		✓
Community Maze	54	max 20	Colored tape. 7x8' tarp, candy bars	35-45	✓	✓	✓		
Dog Biscuits & See Saws	125	teams of 3	One 12" ruler, manila folder, 4" dog biscuit (or similar item for a fulcrum) per team and 15 miscellaneous small objects per person	45-60		✓	✓		
Five Easy Pieces	50	teams of 5	1 set of precut paper sheets per team	30-60	✓		✓		
Frames	132		(Optional: 8.5"x11" piece of paper per person)	5-30	✓				

Activity	No.	Size	Materials	Time				
Group Juggle	237	15-20	1 soft, throwable objects per person, bucket	15-60	✓	✓	✓	
Hands Down	68		5-7 items of equal length	5-15	✓	✓	✓	
Harvest	196	4-40	Container, 6 cups, 200 coins, briefing charts on easel	15-30	✓	✓	✓	✓
Living Loops	179	6-12	One label on a 3' loop per person, ball	20	✓	✓	✓	
Mind Grooving	15		Flip chart or chalk board, pens and paper	5	✓	✓		✓
Monologue / Dialogue	170	4	Two identical line images, flip chart and easel or blank overhead slides and projector	20	✓	✓	✓	
Moon Ball	83	8-30	One large beach ball or balloon	12-20	✓	✓	✓	
Paper Fold	158	5-15	Sheet, or one napkin or square of paper towel per person	5-15	✓	✓	✓	
Paper Tear	91	10-50	3 sheets of 8.5"x11" paper per person	5-10	✓	✓	✓	
Postcard Stories	146	Min 3	A set of 2-3 images per person, all different	5-60	✓	✓	✓	
Space for Living	227	Min 10	One rope loop per person between 2' and 6' in diameter	15-30	✓	✓	✓	✓
Squaring the Circle	114	8-30	A rope 30' or longer	20-30	✓	✓	✓	✓

Game	pg	# of people	equipment needed	time (min.)	mental models	team learning	systems thinking	shared vision	personal mastery
Teeter Totter	61	2-12	2"10' long 2"x6" boards, cement block, 4 eggs, surveyor's tape, plastic wrap	45	✓	✓	✓		
Thumb Wrestling	30		None	10-20	✓	✓		✓	
Toothpick Teaser	45		6 toothpicks or match sticks per person	7-12	✓				✓
Touch Base	109	20-50	60'-90' of rope in a circle on the ground, frisbee, paper plate, or other 1' target	15-30	✓	✓			
Triangles	205	10-50	Flip chart and easel, one numbered label per participant	20-30	✓	✓	✓		
Warped Juggle	38	6-20	3 tossable objects	20-45	✓	✓	✓		
Web of Life	75	8-12	Ball of yarn or roll of surveyor's tape	10-15	✓	✓	✓		

MENTAL MODELS · TEAM LEARNING · SYSTEMS THINKING · SHARED VISION · PERSONAL MASTERY

Mind Grooving

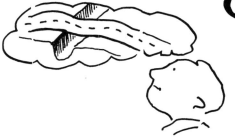

"Our life is what our thoughts make it."

Marcus Aurelius, Meditations

John Wood, founder and president of the Center for Developmental Organizations, reminds us that "unless the thinking involved in a system is developed or evolved, the underlying structure, including the brain, remains unchanged." Part of the challenge of thinking systemically is to vigilantly pay attention to our habitual patterns of thinking.

There are three exercises under the title of "**Mind Grooving**." Each provides an opportunity for people to experience, reflect on and explore their own thinking processes. The first in this series of exercises highlights the effect of socialization on our assumption-making process. The second effectively and humorously illustrates our natural tendency to "lump together" seemingly similar concepts and objects. The third playfully shows how a familiar structure or rhythm can lull our minds into mindlessness. These and many other exercises can work wonders as mental massagers, stimulating experiences to help increase awareness of our own thinking processes.

⇨ To raise awareness of habitual patterns of thinking

⇨ To become aware of how the grooves in our brains impact reflective thinking

⇨ To focus on developing the observer in ourselves so we more often have our thoughts as opposed to "being had" by them

☆ To better understand our own thinking processes

☆ To reflect on how often automatic thought processes can obstruct learning, communication and systems thinking

I like to think of these as "back-pocket" exercises. They can be easily pulled out during a discussion on systems thinking or learning in general to shift a group's attention to its own thinking. I also use them as "ramp-up" exercises to experientially explore the various characteristics of our mental models (automatic, transparent, rapid, etc.) Following a series of these exercises I might launch into a full discussion on the discipline of mental models.

Use these exercises sparingly. Too many in one session can make the facilitator look like the Cheshire Cat with that big, know-it-all grin.

To Run These Exercises

Any number

Approximately 5 minutes per exercise (excluding discussion time)

An overhead or flip charts and marker. Pens or pencils for participants

People will need to be sitting so they are able to see the facilitator and either the flip chart or the overhead projector screen

set-up There is little set-up necessary for these exercises. You may want to arrange for a flip chart or an overhead projector

EXERCISE I: COLOR, FURNITURE, FLOWER

Instructions

1. 2. 3. 4.

Step 1: Participants write on an index card or piece of paper the first word that comes to their minds when they hear the following words:

> color
> furniture
> flower

Step 2: Ask the group how many said "red" for the color.

> How many said "blue"?
> For furniture: How many said "chair"? "Couch"?
> For flower: How many said "rose"? "Daisy"?

With uncanny consistency the majority of the group will have written down red, chair and rose (or one of the second choices). Ask the group why they think this happens.

Debrief

In the West particularly, we pride ourselves on our uniqueness, creativity and individualism, yet socialization is stronger than we realize. There is a physiological reason for this that has to do with neurological pathways in our brains. They can be called ruts and grooves, but a biologist would call them "neural networks." The more we think in a particular way, the deeper the rut we create. When we unconsciously continue in the same thought patterns, these grooves deepen as we reinforce those patterns. The cycle is a vicious one. The more the grooves deepen, the more

---

things look to us as if they fit our groove. Considering that there can be an underlying, natural biological explanation that can enhance or hinder our thinking is a very powerful step toward understanding and challenging our habitual patterns of thought.

This exercise helps us to see that those who did not give the typical responses may be the most potent in helping us to look outside our own mental models. Therefore, the secondary point here is that when it comes to surfacing, testing and exploring our mental models of how the world works, we can be each other's greatest assets. Perhaps, we may want to look for those who disconfirm our current mental models as they may be our greatest source of insight and learning.

Can we catch ourselves going "on automatic pilot"?

How can we encourage diverse perspectives in order to surface, test and explore our own mental models?

Inspiration: Daniel Kim, Stephanie Ryan

Exercise II: EVERYTHING BUT "SLEEP"

This exercise reminds us of the brain's ability to make lightening-speed associations, which at times can be based on fast, yet erroneous assumptions. I would recommend using this exercise in conjunction with the Color-Flower-Furniture exercise and other visually or ented exercises as a way to launch into a discussion about the characteristics of our mental models (i.e. quickly forming associations).

Instructions

1. 2. 3. 4.

Step I: On an overhead or flip chart, show the following ten words:

Slumber	Pillow
Dream	Night
Bed	Blanket
Quiet	Pajamas
Nap	Snooze

Step 2: As you will notice, they are all associated with that life necessity, "sleep." Do not, as the facilitator, draw attention to this. Instruct the group to look at the words but not to write anything down for the moment. After ten seconds, turn the overhead off and ask the group to write down as many words as they can remember, without talking.

Step 3: Ask participants to raise their hands if they wrote down the word slumber. Then ask who wrote down the word night. Then, "Okay, who wrote down the word sleep?" Note how many said they saw "sleep." After those people lower their hands, show the slide or flip chart page again. You won't have to say much...those who thought they saw "sleep" will quickly see that it is not part of the list. Groans, laughs and rolled eyes frequently ensue.

Debrief

You might begin by asking "What happened?" The point here is a simple one:

> How do we develop the observer in ourselves so we more often have our thoughts as opposed to being had by them?

> How do we, in real time, become aware of the associations we are making, and check for their appropriateness?

Every time I use this exercise with a group, I am amazed that anywhere from 50% to 80% of the people in the room will raise their hands when asked if they saw "sleep" in the list of words. What is even more interesting to me is the language used by participants: "Is that a different list?" Or "There's a trick in this somehow!" Try to pay attention to this language and feed it back to the group. There is a fertile discussion waiting to happen.

Exercise III: OAK, JOKE, CROAK...

This is a good exercise to have in your back pocket. I have found it works best in small groups of 10 or less. Many people will not fall for the "trap" (good for them!) but the majority usually do. As a facilitator, the idea is not to put on that "gotcha" grin, but rather to laugh with the participants. I like to use this as a springboard for discussing mindlessness and the power of mindfulness to enhance the capacity to think systemically.

Instructions

At a fast pace, ask an individual or a small group the following questions (pause briefly to allow a response):

Q: What do we call the tree that grows from acorns?

    A: Oak

Q: What do we call a funny story?

    A: Joke

Q: What do we call the sound made by a frog?

    A: Croak

Q: What do we call the white of an egg?

    A: Yolk

Debrief

This and other mental massages (the **Mind Grooving** exercises described in this set) are non-threatening and effective entrees into a discussion of single loop versus double loop learning (see Argyris, "Teaching Smart People How To Learn," *Harvard Business Review*, May-June 1991, p.100). In single loop learning, we cycle back and forth between a problem and a solution. In double loop learning we revisit the mental models we hold about the problem and the possible solutions to that problem. **Mind Grooving** exercises can help remind us to consider our mental models before diving into problem solving.

After running through the exercise, here are a few questions to ask:

> What are the trip wires we need to lay in our brains so we default more frequently into the reflective mode?

> How can we make our conceptual habits less transparent?

Note: If you will be using this exercise with a non- English speaking group, you will have to adapt it to ensure that you maintain a rhyming structure in the group's native language.

VOICES FROM THE FIELD

Dennis Meadows recognizes this exercise as a variation on the game, "Simon Says." The facilitator stands in a circle with the group and rapidly issues directions that are reinforced by the facilitator's example. "Touch your hair" (facilitator touches his hair), "touch your ears" (facilitator touches his ears), "clap your hands" (facilitator claps his hands), "touch your nose" — and the facilitator touches his cheek. See how many people catch on. Silly, yes, but effective in stimulating people to higher levels of mindfulness.

Source: Adapted from Ellen Langer's book *Mindfulness*, originally in G.A. Kimble and L. Perlmutter, "The Problem of Volition," *Psychology Review* 77 (1970): 212-218

Arms Crossed

"The problems we have created in the world today will not be solved by the level of thinking that created them."

Albert Einstein

If Einstein wanted to give people a sense of what it would feel like to change the level of their thinking, he might use this exercise. To build on Einstein's wise warning, we must be willing to continually review and sometimes change our habitual patterns of thought in order to be life-long learners. That is a compelling notion but what we often forget is that the process of changing our personal patterns of thought can be uncomfortable and frustrating. This exercise playfully makes that reality discussible.

 ⇨ To encourage participants to look at the often uncomfortable and awkward feelings associated with learning as potential opportunities for insight and development

 ☆ Increased awareness of the self-imposed challenges to changing the way we think

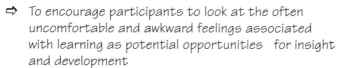

I like to use **Arms Crossed** because it is a wonderful physical analogy to the cognitive process of stepping out of our mental ruts and grooves. This process of recognizing and altering our habitual way of thinking, which often produces our greatest insights and learning, is frequently awkward and uncomfortable. I use **Arms Crossed** to encourage participants to embrace some of the awkwardness as a sign of growth and learning.

To Run This Exercise

 You can do this with any number of people: a single person, a group of 10, or a large group of 300

 The actual exercise takes no more than a minute. The length of discussion is up to you

 Enough to comfortably accommodate participants

 None

 None

Instructions

Step 1: Ask the group to do the following: "Fold your arms the way you would if you were bored, with one arm naturally falling on top of the other. Look at your arms and notice which one is on top. Notice how this feels. Is it comfortable? Does it feel normal?"

Step 2: Now ask the group to uncross their arms and fold them again, the other way, with the other arm on top. "How does that feel? What do you notice?"

Here people may comment that the second way of folding arms feels "uncomfortable," "awkward," or "more alive."

Variation

Clasp your hands together, inter-lacing the fingers naturally. Reclasp the fingers, shifting them over by one finger.

Debrief

I link the physical analogy of feeling uncomfortable when we cross our arms in a nonhabitual manner to the cognitive and emotional experiences we have when we are learning something new. Dawna Markova has often suggested the key question here is:

> How does our need to be comfortable and secure and avoid feeling awkward potentially get in the way of our learning?

It may be that the times of greatest growth occur when we step out of our "comfort zone."

Inspirations: Moshe Feldenkrais, Fred Kofman, and Dawna Markova (author of *No Enemies Within*)

Circles in the Air

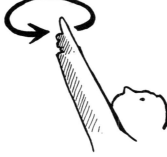

How many times have we heard the lament: "If only those guys up there (in the chairman's office) would get their act together!" Or, "If only the management could see how it really is." We all have a propensity to consider ourselves "outside the system," and to blame someone or something else for the problems we are experiencing.

This exercise works on many levels. It exposes our tendency to see ourselves outside the system and the enemy as "out there." It is also a marvelous springboard for exploring the premise that our particular perspective in a system colors our view of that system. Potentially, if we can change our vantage point either mentally or physically, we may discover new insights and new leverage points.

⇨ To explore the possibility that our viewpoint depends upon where we "sit" and to discover new leverage points in complex systems

⇨ To set a context for discussing the concept of underlying "structure"

☆ Greater awareness of "the enemy is out there" syndrome

This one packs a punch for such a simple exercise. Let's say you're hard at work on a large scale change process with a mixed group of healthcare practitioners — doctors, hospital administrators, nurses, etc. You have just begun a discussion on how the structure of a system creates behavior patterns whose symptoms are what we witness as events. You now want to look more deeply into the level of systemic structures and you want to bring a group's attention to the various perspectives they hold.

As people go through the exercise, they quickly discover that they can simultaneously hold completely different perspectives of the same system (i.e. the pen circles clockwise from one perspective and counter-clockwise from another).

This exercise illustrates how our perspectives affect the actions we take within a particular system.

It subtly focuses a group's attention — in a fun, nonthreatening way — on thinking about its own thinking. See the debrief questions for more detail.

To Run This Exercise

Any number

2 to 10 minutes (depending on length of debrief)

Just enough room to be able to point a pen or finger in the air

A pen, pencil or other straight object

No formal set-up necessary. Participants simply need a pen or pencil. They can be either sitting down or standing up

---

Instructions

Step 1: Ask everyone to pick up a pen (or a pencil).

Step 2: Have them hold the pen straight up in the air, and pretend to draw a circle on the ceiling, in a clockwise direction. Tell them to keep drawing the circle and looking up.

Step 3: Say, "Now slowly continue to draw the circle clockwise, bring the pen down a few inches at a time until it is in front of your face. Continue to circle the pen, and slowly bring it down until you are looking down on top of it. Continue to draw the circle while looking down on it."

Step 4: Ask the group, "What direction is the pen moving?" (It will be a counter-clockwise direction at this point. I smile at those who say "clockwise" and encourage them to try again.)

Note: You will find that some people lose the integrity of the circle as they bring their pens down, swishing their hands back and forth in a straight line. If you notice this, suggest that the person start over and encourage him or her to practice "drawing" a round circle on the ceiling before moving the pen down.

Debrief

The first question to ask is: "So what happened?" The initial responses tend to range from the insightful ("What changed is my perspective") to the self-aware and humorous (see below). After people have had a chance to try it again, most of them will see that what changed as they brought the pen down was not the direction of the pen, but their perspective or vantage point.

The debrief can go in any number of directions. The questions I have found most valuable are:

What was your initial reaction?

What are the first thoughts that came to mind and the first words that came to your mouth?

Do you remember the language you used to describe what happened?

Do your immediate reactions provide any insight into your own process of forming assumptions?

"We don't talk about what we see; we see what we can talk about." I have heard Fred Kofman, an accounting professor at MIT, say this a number of times and now, after hearing hundreds of reactions to this simple exercise, I know what he means. For example, looking in puzzlement at their pen as it circles counter clockwise, I have heard brilliant people say, "my pen is broken" or "you tricked me."

I often wonder when I hear these comments if we may have stumbled onto a language gap.

Have we yet to find the language for the concept of multiple vantage points in complex systems?

Is it possible that changing our vantage point is a way of discovering new leverage points in complex systems? Ask for examples.

This quote from Donella Meadows (a systems dynamicist, author and columnist) can spark a wonderful conversation:

"How is it that one way of seeing the world becomes so widely shared that institutions, technologies, production systems, buildings, cities become shaped around that way of seeing? How do systems create cultures? How do cultures create systems?" (Donella Meadows, *Thinking in Systems* (2008))

In this exercise, how is it that we may all be looking at a system from a clockwise perspective when we could find ways to look at it from multiple perspectives?

VOICES FROM THE FIELD

The reactions from folks who experience this exercise for the first time are delightful and enlightening at the same time. During a session with a group of 40 practitioners of systems thinking, I heard someone call out: "I think my pen must be broken!"

Other enlightening reactions:

"I never did it right in the first place."

"I changed the direction as I brought the pen down."

"Let me do it again and do it right."

"This is a trick!"

It is interesting to see that in the initial reactions to this exercise there is a tendency to blame someone — usually ourselves — for "not doing it right."

Steve Gildersleeve, a management consultant in Canada, called the other day to tell me of his experiences using this exercise: "I recently was working with a group of 300 people and at the end of the presentation I used the **Circles in the Air** exercise. It was really powerful to see the looks of surprise and astonishment. It really worked to spark a conversation around the whole idea of changing perspectives to get a better understanding of complex systems."

MENTAL MODELS

TEAM LEARNING

SYSTEMS THINKING

SHARED VISION

PERSONAL MASTERY

Thumb Wrestling

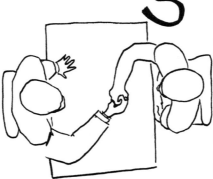

This exercise uses a well-known children's game (thumb wrestling) to provoke rich discussions about collectively held mental models of competition and collaboration. I sometimes worry that I may be having more fun with this one than the group! I like it because it raises awareness of the barriers and enablers to collaborative competition. From my experience, talking abstractly about the properties of mental models in a lecture format is a losing proposition. Eyes glaze over, arms fold, side conversations spring up. But when you engage a group in an experience like **Thumb Wrestling**, through which they can have fun and be students of their own behavior, then you've created a potentially powerful learning experience.

PURPOSE

⇨ To show, in real time, how our mental models (e.g., our deeply ingrained beliefs, myths, stories about how the world works) are often transparent and directly affect the actions we take

⇨ To practice the art of seeing interdependencies and unintended consequences

OUTCOMES

☆ To expose and explore our implicit assumptions about competition and collaboration

☆ A real-time experience of collectively held mental models (e.g., to have to compete to "win")

☆ A context for discussing how our mental models or lenses process the information we take in and act on

CONTEXT

It is one thing to talk about our mental models and another to see them in action. In the case of this exercise, **Thumb Wrestling** gently and humorously exposes our mental models about wrestling, games in general, and more importantly, winning, losing and the potential for win-win situations. My colleagues and I have used this exercise in several ways: to introduce the concept of mental models; as a wonderfully effective practice field for exploring the characteristics of mental models; and as an experiential introduction to conceptual models of thinking processes, such as the "Ladder of Inference" (developed by Chris Argyris, *Overcoming Organizational Defensive Routines*, p.88-89, Prentice Hall, 1990).

To Run This Exercise

Any even number. If there is an odd number, the leader may want to participate

10 to 20 minutes (depending on length of debrief)

None. (Unless you want to give a prize, such as candy, to the winners)

No requirements

Participants sitting in chairs with or without a table

Instructions

1. 🔲 🔲 2. 🔲 3. 🔲 4.

Step 1: Ask participants to find a partner, preferably by turning to the person sitting or standing next to them. If there is an uneven number, the leader may participate.

Step 2: Once everyone is paired, ask the group if they have ever thumb wrestled before. From my experience, more than half have spent long car trips doing this with a sibling in the back seat. Demonstrate for those who don't know what thumb wrestling is. Have the pairs grasp fingers as shown in the following illustration.

Step 3: Explain that the goal is "to collect as many points as you can in one minute." Important: be careful not to set the partners up explicitly as "competitors."

I like to include a first and second prize (i.e. a big and a small bag of M&M candies, especially if I do the exercise in the late afternoon). To get a point, one partner pins the thumb of the other partner (see illustration below).

Step 4: Before beginning, ask each pair to warm up by tapping their thumbs back and forth three times, then when the leader says "go," begin the thumb wrestling.

Step 5: After one minute, stop the game. (There will probably be a lot of laughter and joking, so go with it and have fun.)

Variation

This can also be done as arm wrestling, but beware — it can become quite physical.

Debrief

Ask the partners how many points they've gained. You will hear numbers that tend to hover between one and five, with the occasional pair who manages to get 20 or 30. If you have a pair with a high score, ask how they did it. The answer will most likely be that they cooperated, one person allowing his or her thumb to be pinned by the other multiple times, and then switching. Using this method, the partners have a much better chance of "winning."

My debrief questions are focused on bringing the group through a "what if" exploration: what if we did the same exercise using the lens of a systems thinker?

For example as a systems thinker, we might:

> *Consider mental models:* what were our mental models about **Thumb Wrestling**? Typical answers: one person wins and one person loses.

> *Look for unintended consequences:* in this instance, straight competition creates an unintended consequence: you both lose.

> *Look for interdependencies:* how can we shift our focus to see various forms of interdependence? For example, instead of looking at each other as two adversarial thumb wrestlers, how can we shift our focus to another, higher leverage form of relationship, i.e., collaboration?

Chevreul's Pendulum

Using brightly colored yarn tied to metal washers, an individual or a group can experience the mobilizing power of personal vision and mental models through this exquisitely simple exercise. I especially like to use this in large groups of twenty or more; the more participants, the more powerful the collective "ah ha's."

PURPOSE

⇨ To experience the power of personal vision

⇨ To prepare for work with one or more of the five disciplines, including mental models, personal mastery and shared vision

⇨ To experience the philosophy of personal mastery — that it is more important to hold the vision of what you want than to know how you are going to get there

OUTCOMES

☆ An increased awareness of and ability to create robust personal vision

☆ A common experience from which a group can talk about the necessary steps toward building a shared vision

CONTEXT

Words are sometimes ill-equipped to convey the power, strength and dynamism of a clearly visualized goal or objective. I like **Chevreul's Pendulum** because it allows participants to rely less on language and more on their ability to create a clear picture in their minds of what they want to create, which is, in this case, movement of the washer in a particular direction. This exercise is a fun way of introducing a key component of personal mastery: the life long practice of visualization.

To Run This Exercise

Any number

30 minutes (on average, including debrief)

Participants should be able to prop their elbows up on a table, a desk or a chair. Can be done in circles of 6 or 8, or individually

A metal washer (size of a quarter, one for each participant) tied to a 12 inch brightly colored piece of yarn, paper printed with a "target" (see illustration)

I often prepare the room by placing a set of the necessary equipment on the chair of each participant

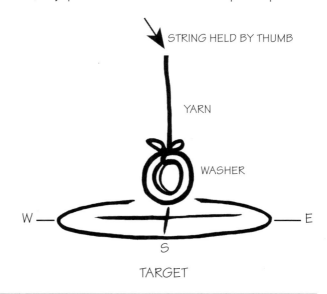

Instructions

Step 1: Each person is given a metal washer tied with one piece of brightly colored yarn (about 12 inches long). The end of the yarn is placed over the thumb and then the elbow is anchored firmly on a desk or table. If there is no table, people can sit on the floor and anchor an elbow on the chair. (see illustration).

Step 2: The washer is hung from the string over the thumb, approximately 1 inch from the center of the diagram. It should not move.

Step 3: When everyone has the washer at a dead-stop over the center of the diagram, ask the participants to: "Picture the pendulum moving up and down in your mind (swinging from the top of the target to the bottom). Do not guide the movement with your hand. Let the picture you have in your mind do the work. Hold that image in your head." After thirty seconds or so, many people will find that the pendulum begins to swing in the direction visualized.

Step 4: Then say: "Use your hand to bring the pendulum to a stop. Now, picture it moving from right to left." Similarly, the pendulum swings from right to left for many.

Debrief

You will, at this point, have a lot of stunned people scratching their heads and looking at you to help them make some kind of sense out of what they just experienced. I begin the discussion by asking how many (by a show of hands) found that the washer moved in the direction they visualized? On average, about three-quarters will have some success. I might then ask the group to consider what force(s) moved the washer? It is through this line of questioning that we can see the connection between our ability to visualize the results we want and actually achieving those results.

Another good question to ask: "What if you were able to tap into that mobilizing power more effectively and frequently?"

Warped Juggle

You wouldn't think that a group of adults tossing koosh balls, stuffed animals and the occasional rubber chicken would add up to a powerful learning experience. This one does and it is tried and true. It provides a real-time experience of common system archetypes, and an opportunity to explore our automatic and often transparent process of making assumptions. After the group has met the challenge posed by the exercise, they have an opportunity to become students of their own behavior as they retrace their actions through group discussion and, if appropriate, causal loop diagramming.

PURPOSE

⇨ To work with one or more systems archetypes, including "Limits to Success"

⇨ To explore the automatic nature of our assumption-making process

⇨ To experience the power of collective mental models

OUTCOMES

☆ Use and examine the creative process for alternate solutions

☆ Draw a loop diagram to map the group's process

☆ Extrapolate to other situations in which exploring assumptions and looking for alternative models are useful or critical

Surface one or more assumptions about team learning and problem-solving

This exercise is particularly good as an entree to the topic of mental models, as it allows participants to discover from experience their own processes of assumption making.

 Traditionally used as a team building exercise, it is also ideal for considering the parallel processes of team problem solving and team learning.

To Run This Exercise

 Min: 6, Max: 20, Ideal: 8 to 12

 20 to 45 minutes (depending on length of debrief)

 Clear away all furniture to create a space large enough for the group to stand shoulder-to-shoulder in a circle. This exercise can be conducted almost anywhere: in a boardroom, on a lawn, in a corridor

 Three tossable objects (i.e. tennis balls, koosh balls, oranges, stuffed animals, rubber chicken) Note: tennis balls can be difficult to catch

 Have the three tossable objects on hand. If possible show only one object at first, hiding the other two in your pockets

Instructions

Step 1: Gather the group into a circle, with you as a participating facilitator. Show one of the objects and begin by tossing it to another member of the circle (but *not* to the person standing next to you). It is *important* to use a gentle underhanded toss. This is not an exercise that should require expert catching skills. Slow the pace of the toss if necessary so everyone is comfortable with tossing and catching the objects.

Step 2: The person receiving the object tosses it to someone else who has yet to touch it. When all members of the group have touched the object, it is tossed back to the facilitator. The sequence is repeated with each person remembering to whom he or she tossed the object and from whom it was received. When the group has sequential tossing of one object down, you can then introduce two more objects to the tossing.

Step 3: The facilitator asks the group to estimate how long it will take to toss all three objects in them sequence the group has established. Before coming to a consensus on the time, you should state that there are only two rules:

- 1) everyone must touch the objects once, and
- 2) they must be touched in the same (human) sequence.

When participants ask for clarification on the rules, it is **important** that you state there are only two (as outlined above). When participants begin to ask how they might "bend" the rules, the two rules should be your standard response. Also, I ask if anyone has done this exercise before. If they have, ask them to participate, but not to offer the solution.

Step 4: Come to a consensus on the time and then, with one of the participants acting as a timer (a digital watch is preferred), try the sequence again. When all three objects are returned to the facilitator, he or she calls "stop" and asks the person with the watch what the time was. Whatever time they end up with (typically the first effort is 20 to 40 seconds), you then challenge them to cut that time in half. (To have some fun, I sometimes spur groups on by saying their major competitor has done it in X seconds less). The exercise is complete when the participants feel they have done it in the fastest time possible, usually in a second or two.

Possible Solution

Group members will figure out that they should stand next to the person to whom they are tossing the object. A shuffling then ensues until each is able to pass the object to the person next to them, rather than tossing it across the room.

Variations

If group members are new to each other, ask each person to call out the name of the person to whom they are throwing the object. The person to whom the object is thrown, receives it, saying, "Thank you, Ann," and then tosses it to the next person, saying his or her name.

You may offer a member or members of the group the role of observer. Another way to phrase this is, "We need a TQM person, any volunteers?" Take this person aside and ask him or her to asses the group's process: what happened when someone had a contrary idea? How did the group solve the problem? What patterns of behavior did you observe?

Debrief

What typically happens is that initial efforts lead to improved performance. Over time (usually within the first 5 to 10 minutes), the group cuts the time down from 40 seconds to 10 or 12 seconds but then they encounter a limit. This limit often causes the performance to slow down or even stop, even though efforts to solve the problem may be increasing. An example of "increasing efforts" might be that the group decides to squeeze in tighter together or to throw the ball faster (which actually causes more errors and more delays). At this juncture, the opportunities are rich for gaining insights into individual and group behavior patterns within complex systems.

One way to do this is through the use of causal loop diagramming. Ask the group to identify the key variables in their experience (e.g., teamwork, time pressure,

improvements, etc.) and begin, using a flip chart
or overhead, to map the relationships between the
variables. Following is a sample diagram.

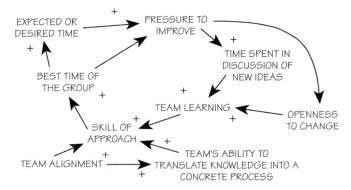

Diagram created by Dennis Meadows

If the group has been exposed to the systems
archetypes, ask if they see any such archetype in their
own problem-solving process. The "Limits to Success"
archetype, for example, typically involves a constraint:

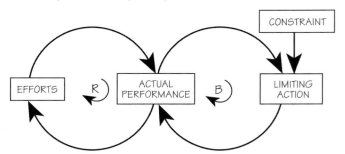

Archetype drawn from *The Fifth Discipline* by Peter Senge
R is a reinforcing loop, B is a balancing loop

Ask what they think the constraints were. In **Warped
Juggle**, the constraint is very often the group's
assumption that there are more "rules" than those
stated by the facilitator. What is the limiting action?

The limiting action here can be that participants
hammer away at the same approach, without stopping
to reflect on their assumptions, hear other ideas, or
consider other options.

Transference to professional and personal experiences: The group experienced how immediate success can produce subtle constraints, particularly in the thinking of individuals and groups. You might ask: "What kinds of inherent pressures and constraints are accumulating in your organization as a result of its success?"

As a facilitator, you can also point out that the way in which we receive information affects the assumptions we make about that information. In this exercise, the facilitator begins by tossing the ball across the circle. Participants assume that they too have to toss the ball, even though there are no requirements in the rules to do so. The fastest times are actually achieved by *not* tossing the objects.

VOICES FROM THE FIELD

Andy Bryner, a friend and colleague who is a master at creating physical learning practices, used **Warped Juggle** recently with his partner Dawna Markova. They were working with a group of healthcare administrators to "develop an awareness of the whole, and all the variables which influence the whole." Andy and Dawna facilitated eight groups of fifteen people each (a feat in itself!) and had the following experience:

"Even though the two rules were spoken and written, one group continued for most of the allotted time tossing the ball as had been demonstrated in the beginning. They improved greatly over time and had a lot of fun and they never redesigned their structure to meet the constraints in a more efficient way as did other groups. In the debrief, they owned that was true of their unit at work. In service they experienced great team spirit and enjoyment and not a lot of innovation, examining of mental models or rethinking processes.

"Another group immediately understood that their structure could be redesigned, tried the first way, and then spent up until the very last minute planning, and managed to accomplish the task in five seconds. But even this great time* brought some discomfort with the process: there were only a few vocal planners, lots of ideas were disregarded, and there was not much experiential learning. Many felt out of the creative loop. In the debrief, they talked about how in their unit there were a few super planners and many quiet complaining "compliers" which, over time, produced withholding of resources and dependency on a few. Considering systems thinking, the group talked of an awareness that short term success may actually have the unintended consequence of blocking future learning and greater effectiveness."

*The best time is often under one second—LBS

Toothpick Teaser

The **Toothpick Teaser** exercise helps us to explore a universal phenomenon: when given data, whether it is a symptom of a problem to be solved or a schedule to be adjusted, the way the data is presented to us affects the possible questions we ask and solutions we see. This is true unless, as Diane Corey reminds us, "we are highly conscious of our own mental models and assumptions." More than a simple "thinking-out-of-the-box" activity, this exercise helps us to collectively reflect on our instinctive approaches to problem definition and problem solving.

⇨ To encourage participants to look at all of the factors influencing their ability to learn and solve problems, especially the means by which a challenge or problem is presented

☆ A mental massage, stretching our brains to think beyond our current mental models

☆ An improved understanding of personal problem solving approaches

---

☆ An increased awareness of the power of examining the manner in which data is presented prior to problem solving

To some, this will look and feel like a traditional brain teaser, so be ready for a few groans. I usually have a good laugh with the group and note that later we might talk about those groans, which are a good source for mental model exploration.

I often find myself pulling out the box of toothpicks when I want to make the connection between examining mental models and improved problem definition and problem solving. An excerpt of my conversation with a group might sound something like this:

> "It's fair to say that we all solve problems from certain understandings and past experiences. The problems themselves are often not complex and there are many tools out there to help 'problem solve.' The complex issues are our understandings, or our mental models. And what we often forget to do in terms of problem solving is to go back and reflect on our original understandings."

When we cycle back and forth between problem/solution, we are on what Daniel Kim calls "the problem solving treadmill":

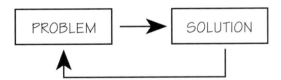

In the toothpick exercise, most of us immediately launch into solving the problem (I did the first time), without considering the mental models we have about the problem or the way in which the problem was presented to us. I consider this a mini-practice field in which we can practice the life long art of consistently reflecting on our mental models.

To Run This Exercise

 Number of People Any number will work. With large groups (15 or more) you may want to have participants work in pairs

 Time Approximately 2 minutes to explain; 5 to 10 minutes to do

 space Floor or table upon which to place the toothpicks

 Equipment 6 toothpicks or match sticks per person

 set-up If you can set up the room in advance, put 6 tooth-picks flat on the table in front of each person

Instructions

1. 2. 3. 4.

If advance preparation wasn't possible, place a box of toothpicks within reach of each person. Ask participants to each take 6 toothpicks and place them flat on the table. Using all 6 toothpicks, ask them to cre- ate four equal sided triangles.

Possible Solution

One solution requires the person to "think outside the box" and to break out of the one dimensional mode. Lay three toothpicks flat on the table to form one triangle. Use the remaining three toothpicks to create three new triangles by building a teepee-like structure.

Debrief

Participants are, in a way, "set up" because I have them place the toothpicks flat on the table in front of them. The solution requires them to think in 3-D. Part of the obstacle becomes the way the challenge is presented. (**Warped Juggle** requires a similar thinking process.)

Some questions I ask:

How did I "set you up" when I instructed you to put the tooth picks *flat* on the table?

If you had a partner, in what way did he or she encourage or discourage "out of the box" thinking?

VOICES FROM THE FIELD

Diane Corey, a story teller and organizational learning educator, is masterful in her integration of experiential exercises to illustrate key concepts of Organizational Learning. Here Diane speaks from her many years of experience using this exercise:

"I ask people to work in pairs, and line the toothpicks up like a picket fence on a flat surface (often a notebook lying on their knees). To illustrate, I put the toothpicks flat on an overhead, lined up next to each other. Then I give them the directions and after two to three minutes I'll ask if they're ready for a hint."

At this point Diane explains that the problem is much easier to solve when someone gives you a coaching tip. In the case of the toothpick exercise, her coaching tip is: "You have to solve it two and three dimensionally."

After that, most people groan and solve the problem right away.

In her debrief, Diane asks:

"How did I set you up not to be able to solve this puzzle? How does this relate to your personal and professional lives?"

Through the exercise, Diane helps the participants to consider a universal phenomenon: any time we are being given data, the way the data is presented predetermines the outcomes and possibilities we see, unless we are highly conscious of our own mental models and assumptions.

Before the exercise, Diane may use visual brain teasers to bring the group's attention to different visual paradigms (W. E. Hill's old woman/young woman is a good example). Her question to the group is "How would you coach someone so that they could see the image differently? To coach someone, you have to find the good questions that allow another person's perspective to change." Diane relates this to the skill of balancing advocacy and inquiry. She urges the group to talk about how we can continually help each other shift our focus and consider how that process would look in a meeting.

Following the toothpick exercise, Diane often has the group look through newspapers or company documents and highlight mental models.

For those who will be trying this exercise with groups, Diane reminds us, "All of these exercises are more powerful if they are part of a flow — a thoughtful integrated structure."

Source: "Problem Solving" 1963 M. Scheerer: Scientific American 208: 118-28. With inspiration from Diane Corey.

MENTAL MODELS TEAM LEARNING SYSTEMS THINKING SHARED VISION PERSONAL MASTERY

5 Easy Pieces
(or The Schein Shuffle)

As we see in our everyday lives, the basic pattern of life is a network of interconnected systems. Within a community, for example, there are many sets of interconnected systems: education, business, social service, religious organizations, healthcare, etc. Yet often under the pressure of time and every day life, we act as isolated, disconnected units. The author and physicist Fritzjof Capra reminds us that the first principle of ecology is interdependence. How can we develop the habit of mind to be attuned to this principle in our everyday lives?

This exercise is unpretentious, slightly disarming and ideal for illustrating interdependence, an awareness of which is vital to the development and practice of systems thinking.

⇨ To experience a shift in perception from object (the set of cut-up pieces) to relationships (among the team)

⇨ To explore our knee-jerk tendency to "go it alone"

☆ A greater ability to identify mental models in real time, and see key inter-relationships and systemic structures

This exercise takes some advance planning, so I usually use it when I'm working with a group for a day or more. I like using **Five Easy Pieces** to jump start a conversation about the "Ways of a System Thinker" (see **Guiding Ideas**).

To Run This Exercise

You will need a minimum of 5 people and then any additional multiple of 5

TIME The exercise itself should take no more than 20 minutes. The debrief, when related to similar organizational experiences, can take about a half hour

 Space Enough for 5 people to sit in a circle in chairs at a table, or on the floor without a table

 Equipment 5 pieces of 10 inch x 10 inch colored paper or cardboard; scissors; ruler; a pencil

 Set-up Prepare the pieces: for each group of 5, cut up five 10 inch x 10 inch pieces of colored paper (card board is preferable, or something that you can laminate). Cut the shapes as described below. The numbers are to guide you in the cutting process (same number, same shape) but the pieces used by the participants should not show a number. Once the shapes are cut, mix them up and divide the pieces into five piles, with three pieces in each.

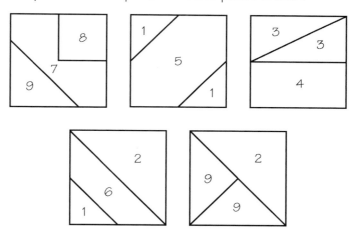

Instructions

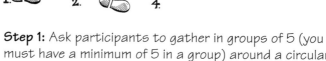

Step 1: Ask participants to gather in groups of 5 (you must have a minimum of 5 in a group) around a circular table or in a circle on the floor. If, for example, you have 50 people you can either divide them into 10 groups of 5, or 5 groups consisting of 5 pairs.

Step 2: Give every person or pair three random pieces.

Step 3: Participants are told the objective of the exercise: "Each team member is to form a square (flat on the table or ground) with the cut up pieces of paper." Be careful not to suggest that the members of the team are competing with one another. A participant (or pair) wanting to exchange a piece can place it in the center of the circle. A piece placed in the center may be taken by another participant. Participants should hold no more than three pieces at one time.

Special Rules

No talking

No folding of paper pieces

No overlapping of pieces

All pieces should be used (each square is comprised of 3 pieces)

Exchange 1 piece at a time

Debrief

Most often, unless someone on the team has played this game before, the first reaction will be for each person to try to solve the puzzle individually. Eventually, someone (or a pair) will either sit smugly with a square in front of them or raise their hands and say, "I got it!" At this point, you may have to remind the group of the objective: each team member is to form a square with the cut up pieces of paper. Therefore, one complete square is not enough. There must be five complete squares in the group.

The person who completed a square may have to give up a piece of that square so all members are able to form squares. This exchange seems counter-intuitive to many at first.

As usual, I ask the simple question: "What happened?" and then let the story unfold as the group experienced it. The key points to touch on in this exercise are:

How, in real-time, we can shift our perceptions from objects (the cut-up pieces of paper) to relationships (among the players)?

How our knee-jerk tendencies to "go it alone" can create barriers to effectively seeing the interdependencies in systems and inhibit problem-solving.

How a greater ability to identify mental models enables us to more readily see underlying systemic structures and key interrelationships.

Inspiration: Edgar Schein, U.S. Armed Forces

Community Maze

The **Community Maze** is a long-standing favorite in the world of experiential education. It is also known as the "Corporate Maze." I have found it to be a terrific practice field in which our mental models readily surface, so that our tacit and implicit beliefs can be challenged and enriched.

It is also a wonderful tool for exploding myths associated with teamwork, such as "one and one make three." Most of us who have participated in teams know that many teams end up operating below the level of the lowest ability in the group. "The result is skilled incompetence," says Peter Senge. People in groups grow incredibly efficient at keeping themselves from learning.

PURPOSE

⇨ To explore the discipline of team learning: tapping into the emerging collective intelligence (and memory) of a group

⇨ To look at the cost of missing the "win-win" opportunities and the benefit of "collaborative competition"

⇨ To explore the interdependent nature of complex systems

OUTCOMES

☆ Experience one or more barriers to team learning. For example, under-utilized interpersonal and group inquiry skills

☆ A real-time experience of how inadequate feedback in a system hinders the learning process

Community Maze is an ideal exercise for intact teams or work groups. It is especially relevant for those working in matrixed organizations or for those who are being asked to exchange information and knowledge with multiple groups in an organization.

To Run This Exercise

 Maximum 20. If there are 12 or more participants, split them into two groups. Position the groups at opposite ends of the maze

 35 to 45 minutes on average (depending on length of debrief)

 Enough for the group to stand in a loose circle around the maze

 Masking tape, duct tape, or any brightly colored tape. To create a portable grid, you'll also need a 7 foot x 8 foot piece of plastic or tarp. Enough candy to represent $1 million for each team (I usually give each group ten "$100,000" candy bars)

 Follow the instructions to create the grid, leaving ample room for participants to gather around its sides. Each facilitator should have a chart of the maze, a pen or pencil and a noise-maker (a kazoo, for example)

Instructions

Step 1: Lay out the grid on the floor. Outside, you can use brightly colored surveyor's tape and small stakes. You can also make a portable grid using 7 foot x 8 foot plastic or tarp and duct tape.

Step 2: Use the tape to create 8 squares across, 7 squares down with each square measuring 1 foot. If you are using a conference room, the tape can be quickly applied directly to the floor.

Sample Framing of The Community Maze

For this exercise be careful not to suggest that the two groups are two opposing teams. Rather, let them know that they are two units of the same company, two countries sharing the same water supply, two departments in a non-profit organization, etc.

There are many ways to frame this experience: isomorphic, universal and fantastical are just three. I usually like to set this one up isomorphically, making it as relevant to the participants' real-life organizational situations as possible. Here is a sample framing for a corporate group followed by the complete set of instructions:

"You," (addressed to the two groups standing at both ends of the grid) "are two business units of the same corporation (marketing and research, for example). You have been given a budget of $2 million ($1 million for each unit). Your objective is to navigate your way through the grid, which represents your newly matrixed organization, while spending the least amount of your limited resources.

"The grid is full of obstacles, pitfalls and opportunities. There is only one right and safe path, which I know (because I have the map)." (Both teams travel the same path, but they should not be told this. If asked, be non-committal and say, "They could be.") "Your job is to find the right path by trial and error and get all the group members to the other side."

Special Rules

1. You have 5 minutes to plan and 20 minutes to silently attempt the exercise. After the 5 minutes of planning, your team must not talk. Each time a team member speaks during the 20 minutes of silence, the cost is $100,000. The team may purchase an additional 5 minutes of planning (talking) time for $100,000.

2. You must decide in advance the sequence in which group members will attempt to find the path. You must stick to that sequence.

3. One member from each group is allowed on the maze at a time. After a try, that person leaves the grid and the next person in the sequence steps on.

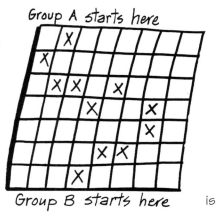

Group A starts here

Group B starts here

4. You may not step over a square. Any contiguous move acceptable. is

5. If you step in an incorrect square, the facilitator will use the noise-maker and you must return the same way you entered. If you do not, that is considered an "avoidable mistake."

6. You have $1 million as your budget. (I usually give each group ten "$100,000 candy bars). An avoidable mistake will cost you $100,000. A mistake is stepping on an invalid square that has previously been stepped on OR not leaving the maze on the same path by which you entered.

7. No "Hansel and Gretel." You may not leave any kind of visual trail or make a map by any other physical or visual means. If you are not the active member attempting the maze, you may not touch the maze. Gesturing wildly is acceptable.

8. You can purchase additional planning time for $100,000 per minute.

Using two groups requires two facilitators (or one facilitator and one helper). Each of the two should be holding a copy of the grid with the correct path marked on it. Obviously, participants are not allowed to see this paper.

Facilitator A with group A holds the map right-side up, Facilitator B with group B holds the map upside down. You can create any path you like but make sure there is only one possible path from square to square. I like to make it look somewhat like an "S" so the group is forced to go backwards at one point, which is quite counter-intuitive for most.

Debrief

By the time you get to the debrief you will have a lot of excited people generally enjoying their ability to talk again after being mute for the past 20 minutes. As a facilitator, I like to ride with this energy and begin the debrief by asking a simple question: "What happened?"

You'll hear many different versions of what happened — from the person who could conceptualize what was going on but was frustrated that he or she couldn't transfer that knowledge to the rest of the group — to the person who wanted to exchange information with the other group but felt thwarted by either group pressure or lack of time.

After multiple stories have been told about what happened on the maze you might then pull a flip chart over and ask the group to consider the various perspectives and levels of understanding they hold of their common experience:

> How does the structure of the maze (the physical make-up of the grid, the rules of the game, etc.) drive the behaviors exhibited by the players?

> What patterns of behavior or events occurred during the game?

Were group members in a reactive mode?

Help the group to practice moving up the ladder, beginning with the events (by describing what happened) to the patterns of behavior over time, and eventually to the structure. This can be an effective, experiential way to introduce "The Vision Deployment Matrix" (Daniel Kim, *The Systems Thinker*, 1995.)

INCREASING LEVERAGE

STRUCTURE

↓

PATTERNS OF BEHAVIOR

↓

EVENTS

Using Systems Archetypes

Systems archetypes are helpful conceptual tools, made up of causal loop diagrams which promote discussion and understanding of stories about interrelated systems. Groups with a good understanding of the basic principles of systems thinking may consider the "Fixes That Fail" archetype as a lens through which to look at this exercise (see "Toolbox Reprint Series," *Systems Thinking Tools*, by Daniel Kim, p.20).

The nature of the "Fixes That Fail" phenomenon is one in which "the squeaky wheel gets the grease." For some groups, this archetype is a useful lens through which to look at their experience on the maze: the person on the maze who gestures and grabs the attention of his or her

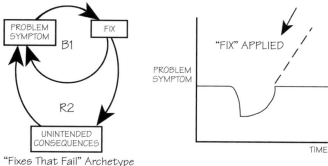

"Fixes That Fail" Archetype

cohorts is the one who will be attended to, presumably then making fewer mistakes. However, what happens (when there are two groups on the maze) is that one group gets distracted and falls into a reactive mode, paying less attention to their own path and more to the other group's path. The unintended consequence is often an increase in the number of avoidable mistakes and a failure to make fundamental changes in the way the group solves problems, communicates and learns together.

Barriers to Learning

When thinking and talking about learning in organizations, the **Community Maze** works well to illustrate a theory expressed by John Sterman (MIT Professor of Systems Dynamics): "All learning depends on feedback." For learning to take place, every link in the feedback loop of learning must work effectively. I often ask a group to consider the following learning cycle:

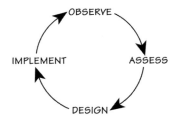

The group begins by observing the maze and assessing the problem to be solved. They then attempt to design a process (e.g. line people up on the side of the maze to remember which squares are the "good" squares), and after five minutes launch into implementation. In the **Community Maze**, participants go through one round of the learning cycle, but eventually get stuck in the implementation stage. They typically do not take the option of buying additional time to plan and so find themselves unable to cycle through again to discover and transfer essential learnings to the group.

Reference: Original version of game instructions from Bill Abelow and New York City Outward Bound Center Professional Development Program staff

Teeter Totter

Teetering over two raw eggs is such an unusual way to learn about and experience the concept of delays in systems. Yes, but that's what makes this exercise effective — a group of people physically experence the concept they are trying to learn. The **Teeter Totter** also gives a group a concrete experience in team learning, an opportunity to explore different leadership styles, and permission to have a generous amount of serious fun.

PURPOSE

⇨ To experience the concept of delays and accumulations in complex systems

⇨ To explore barriers to learning in complex systems, particularly - Inadequate and ambiguous feedback, and - Consistent misinterpretation of the feedback

OUTCOMES

☆ Acknowledge the reality that in complex systems, blame is often misplaced

☆ Challenge linear thinking and encourage exploration of multiple influences and causes

☆ Look at the system (in this case, the teeter totter and the team), and realize problems emerge from within the system as a result of the interaction of the various parts, not from some external event

☆ Gain a different understanding of the barriers to learning in complex systems

This exercise is particularly good as an opener for both mental model and systems thinking work, as it allows participants to discover from experience their own processes of assumption making. At a novice level, the discussion following the exercise can readily focus on time delays and oscillations in systems. An intermediate level of discussion can be had using causal loop diagrams.

To Run This Exercise

This depends on the length of the board you use. With a 10 foot board 11 people can participate, 12 if you squeeze. Any more than 12 on the board is unsafe. (Others can act as "TQM" observers)

Approximately 45 minutes

Clear away all furniture. Use an area large enough to accommodate a wide circle of participants around the teeter totter

Two 10 foot 2 x 6 boards, roll of duct tape, surveyor's tape, four raw eggs, plastic wrap, cinder block

To make the teeter totter, place one 10 foot long, 2 x 6 board on top of the other. Screw the boards together at the ends, then duct tape over the screws and around the ends for extra safety and strength. Set the cinder block on its side and lay the boards on top (see illustration) to create a see-saw. Place brightly colored surveyor's tape on the floor in the shape of a "V," with the point at the center of the concrete block, so you've created a pathway that funnels participants to the teeter totter.

Place a whole, raw egg under each end of the teeter totter. If you are conducting this exercise inside, you might want to place plastic wrap underneath each of the eggs. (They may be crushed by the teeter totter!)

Safety

You and any team members not participating should stand a few feet back from the teeter totter to "spot" the group members. This means you are ready (with your hands up and your knees slightly bent) to steady anyone who may get off balance. Participants should wear flat shoes; do not allow high heels or bare feet due to the potential for a twisted ankle or splinters.

Framing

If the group is comprised of manufacturers or marketers, I might frame the exercise isomorphically, saying: "Your challenge is to balance the twin devils of liquidity and credit risk" as I place the two raw eggs underneath each end of the teeter totter. Or, if I want to tie into the later discussion of an underlying causal loop, I might substitute "supply and demand" for liquidity and credit risk.

Instructions

Step 1: After framing the exercise, ask the group to assemble inside the "V." Their task is to get all members on the board without stepping out of the "V" or on the cinder block. (You can make up your own penalties for stepping on any ground outside of the "V." My favorites are blindfolding a team member, or muting whoever is the leader.)

Step 2: They have 20 minutes to prepare and to get on and off the teeter totter without, of course, breaking the eggs. They can help each other on and off the teeter totter as long as they are standing inside the "V." Carrying or lifting team members in any way is not allowed.

Step 3: Before getting off, they must sing one round of a simple song of your, or their, choice (such as *Row Row Row Your Boat*).

Variation

If you have a particularly precocious group who does this exercise without a hitch, offer them an advanced challenge: doing it again, but blindfolded or mute. If you opt for this variation, you will want to stay very close to monitor the team as they get on and on off the teeter totter and as they stand on the board. Enlist an observer to help you.

Debrief

Traditionally used as a team building exercise, I've found **Teeter Totter** to be effective for teams involved in on-going organizational learning efforts. I've run this exercise at least 100 times now, and approximately 80% of the time the following occurs:

The group successfully gets all members on the teeter totter and joyfully sings one round of *Row Row Row Your Boat*. They did it. And within the time limit.

Then someone remembers that the instructions were to "get on and off the teeter totter in 20 minutes without breaking the eggs." They then proceed to swiftly get off, moving at a faster pace than they did getting on. As each pair steps off, the teeter totter begins to oscillate slightly. Those who are getting off do not pay much attention to this because they are focused on the task at hand — getting off the teeter totter. Under a time pressure (which I make sure to repeat), the group moves even faster. Paying less and less attention to the subtle signals of the teeter totter, the last person or pair on the board is left with a whip-lash effect, eventually tipping to one side or the other and breaking the egg. There's a whoop from the participants and they rush to straighten the board, only to have it go down on the other side and break the second egg.

The team is dejected and disappointed. But not as dejected as whoever was left standing on the board. In the debrief, I find out how that person (or persons) felt when the eggs were broken. One woman told the group, "I feel terrible. I let the team down." In fact, as we discovered through our conversation, she hadn't let the group down. The momentum had begun to build long before it was her turn to step off the board. If an event like this does occur, it can be a terrific segue to a discussion about hidden time delays between a cause and its effect and an appreciation of the notion that in complex systems, there is no blame.

As with most debriefs, I tend to begin by asking the question "What happened?" The team will describe the event as they saw it happen. You may want to help the group elevate their stories up the "Levels of Understanding in Systems" ladder (as described in the **Community Maze** exercise).

Here are a few questions I ask:

> What dynamics of complex systems, if any, did you observe or experience?

> Were there any unintended consequences to your actions or behaviors?

> Can you create a causal loop to represent this system? What feedback loops did you experience?

Once the group has described the event ("We got on fine, but then we lost control." Or "John broke the egg.") encourage the group to raise their stories to the patterns of behavior level. They might say, "Juanita emerged as a leader and we followed her, but no one would listen when I said the board was shifting." Or "Someone at the end kept shifting and making me feel that the board was getting more and more unbalanced," etc. Then move on to describing the hard and soft structures of the **Teeter Totter** system (the physical structure, the rules of the game, group's mental models, etc.).

At this point, you've spent about forty minutes with this exercise (including a basic debrief). Depending on the sophistication of the group and the amount of time you have, you may want to progress to a working session using causal loop diagrams.

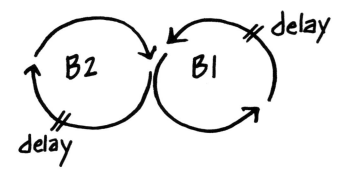

Working with Causal Loops

We see in the **Teeter Totter** exercise a physical example of the systems archetype known as "Balancing Loop with Delays: Teeter Tottering on Seesaws" (Daniel Kim, Systems Archetypes, 1993).

The resemblance is more than just physical. Let's think back to the goal of this exercise: to keep the team members and the board in a state of balance (much like the goal of the marketplace, which is to keep production and consumption in balance). Similar to marketplace dynamics, the adjustments made during this exercise in an attempt to stabilize the system often result in "overshoot and collapse" (or the smashing of a raw egg). This is due to the presence of delays experienced between making an adjustment on the board and the time it takes for another team member to feel it at the other end. In the case of the marketplace, the balancing loops try to stabilize on a particular price. But as Daniel Kim points out in his description of this systems archetype, "The process is complicated by the presence of significant delays."

As a facilitator, you many want to ask the group to relook at the **Teeter Totter** and the team through the lens of a balancing loop, encouraging them to draw the loop themselves, and noting the time delays that occurred.*

With the debrief and causal loops fresh in their minds, you might want to give the group another chance at the **Teeter Totter**, if time permits.

*About the balancing loop with delays: With the addition of delays in this basic feedback loop, the structure becomes complex because the delays often make the resulting behavior quite unpredictable. As Kim wisely points out, "The delays in a typical system are rarely consistent or well-known in advance, and the cumulative effects are usually beyond the control of any one person or firm." As you'll see, this is just the experience you will have on the teeter totter.

Inspiration: Michael Crehore and The New York City Outward Bound Center

Hands Down

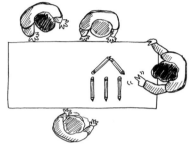

"We do not describe the world we see,
we see the world we can describe."

René Descartes

From our brain's perspective, life is a staggering onslaught of completely disorganized stimuli. Impressions, feelings, images, sounds, smells, bombard our brain through our eyes, ears, nose, and skin. It's confusing, never-ending, and it could be hopelessly chaotic, if our brains didn't have a system to file away all of the information received. In an old Superman comic, an arch villain tries to "do-in" Superman by radically amplifying the hero's senses. Suddenly he hears every noise, smells every smell, etc. The in-rush of sensory data is too much and Superman teeters on the edge of madness.

The process we use to deal with these data is mostly unconscious, multi-layered and highly complex. But it is that very skillful, automatic and rapid process that sometimes causes errors, stereotyping, and inaccurate assumptions. As systems thinkers, the better our mental models match current reality, the more accurately we can understand and intervene in complex systems.

⇨ To raise awareness of the lightening speed with which we make assumptions

⇨ To encourage participants to slow down and examine every assumption, especially the unconscious ones

⇨ To explore the double edged nature of those mental models. They enable us to function in the world, but also often act as blinders.

☆ An embodied experience of using "going wide" or using peripheral vision

☆ A mental climb down the "Ladder of Inference" to "observable data"

☆ An awareness of several mental model characteristics, including: rapidly-forming, transparent, resilient

We rarely plan on using this exercise. It always just comes to mind when the conversation moves to the paradoxical nature of mental models. On the one hand, the very advantages of rapid, automatic thinking for dealing with routines (like tying your shoes) become disadvantages when we're faced with change and uncertainty. Because filters or cognitive structures are so strong, we often see only what we already know, ignore critical information, and rely on standard routines in challenging situations. **Hands Down** confronts head on our habitual patterns of processing information.

It is also quite a non-threatening and fun way to introduce or reinforce what Chris Argyris calls a "Ladder of Inference"—a common mental pathway of increasing abstraction often leading to misguided beliefs. **Hands Down** in particular provides a mirror on the way we select from observable data and the speed with which we tend to make leaps of abstraction.

resources

For a comprehensive explanation of the Ladder of Inference see *The Fifth Discipline Fieldbook* (pages 242-246) by Peter Senge, with Art Kleiner, Charlotte Roberts, Richard Ross and Bryan Smith. Also, Chris Argyris' *Overcoming Organizational Defenses* (page 87), available at most book stores.

Once we're anchored at the bottom of the Ladder of Inference, the most needed and most often underdeveloped skill is peripheral vision (what Dawna Markova calls "going wide"). The wider our perspective, the more data we can take in, and the more possibilities exist for effective action.

To Run This Exercise

 Any number as long as all participants can see several objects in front of you. About 30 people is comfortable, more and it becomes difficult for people to see

 This can go for 5 minutes or it can go for 15, depending on the size of the group and the extent of the debrief

 Enough room so the group can see you kneeling on the ground or sitting at a table in front of them

 5 - 7 items equal in length (we like pencils because they're handy, but dowels, sticks, forks or knives will do as well)

 In this particular exercise, a minimum amount of set up is best. The more spontaneously you can whip out the pencils or forks and start this exercise, the better

Instructions

Step 1: Kneel on the floor, or sit at a table in clear view of all participants. Put the 5 to 7 objects down in front of you, flat on the floor or table. Be a little fussy, as if you are arranging them in some particular pattern.

Step 2: Ask the group: "What number am I showing?" While asking this question, place one or both hands, palm down, on the ground next to the pencils in such a way that your fingers indicate a number from one to ten. The number of objects you have on the ground has nothing to do with this. The number you want people to see is the number of fingers you have extended. A fist represents zero, all fingers out on both hands represent ten.

Typically the group will guess the number of objects you have down in front of you, rather than the number you are showing with your fingers. If you think someone has guessed the number, ask them to try again in the next round. That person may not be able to guess the number correctly a second or third time. If they do perceive the way in which you are communicating, ask them or signal them to be silent and hold "the answer" until the end.

Arrange several different patterns, so the members of the group have ample opportunity to try and understand how you are communicating. It is important that you do not vary the number of objects you use in creating the patterns. If you start out with 5 objects, stay with that number all the way through the exercise. Changing the number of objects would be confusing.

If no one gets it after 3 to 4 tries, start to wiggle your fingers or nonchalantly drum your fingers on the table or floor. That catches some people's attention. Next place your hand on the ground with a slight slap, drawing attention to your hand.

Debrief

Without a thoughtful debrief, this exercise could easily fall into the "gotcha" category and leave participants feeling frustrated and even manipulated.

Once most of the group understands how you are communicating, you may choose to go in any number of directions with the discussion:

Introduce participants to, and/or explore participants' thinking processes through the Ladder of Inference

Make the connection between systems thinking and "reflection-in-action" (a term used by Donald Schön in his book *The Reflective Practitioner* to describe a particular kind of thinking that enables people to question the assumptions behind their actions)

The Ladder of Inference

Hands Down creates a non-threatening situation in which participants can become aware of themselves climbing up and down the Ladder of Inference— selecting, interpreting, and acting on their own assumptions.

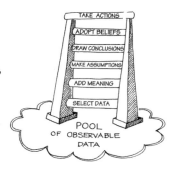

When discussing the Ladder of Inference, we've found this question useful: "If we talk about climbing down the Ladder to 'observable data'—what you would capture on videotape—what would you identify as the pool of observable data in this exercise?"

Reviewing the exercise in their minds, people may say the observable data included my hands, the objects, and my voice. Through this review of mental videos, participants are exploring what it means to "go wide" or use their peripheral vision. Openness to peripheral vision depends on our ability to reject reductionism, to operate with multiple layers of attention and, as Mary Catherine Bateson says, to "savor the vertigo" of doing without easy, immediate answers.

Reflection-in-Action

When discussing "reflection-in-action" you might consider the negative consequences of automatic thinking and the positive consequences of reflective thinking. Our mental models can trap us. This is what quantum physicist Danah Zohar calls, "paradigm paralysis." It's this paralysis that prevents us from seeing beyond the pattern of objects on the table.

Our mental models can be transparent and unconsciously formed, but they can powerfully affect what we see or don't see. Danah gives this example:

"When medieval people went to the seaside, they didn't see the curvature of the horizon. To them, the earth was flat and their mental models simply filtered out the observed data of the curved horizon. When we go to the seaside today, we may perceive the curvature of the horizon because we've been taught in school the earth is round."

Victor Friedman, a senior lecturer at the Ruppin Institute, says, "Negative consequences of automatic thinking have been responsible for failures in interpersonal relations, professional practice, organizational decision making, and modern warfare." Well-known examples of this include The Maginot Line based on the trench warfare paradigm of World War I, the application of Taylor's time-motion studies to industrial process design on the paradigm of the organization as machine, and the current paradigm that "economic growth is intrinsically good" which can lead to such unintended consequences as resource depletion, and environmental degradation.

Reflective thinking, on the other hand, enables people to see their mental models in action, to see situations differently, and to experiment with novel responses. How could reflective thinking help avoid failures in family relationships, organizational environments, government strategies, and modern warfare prevention efforts?

As a quick follow-up to this exercise, we've told the following assumption-teaser (It is wise to test the waters and see how many have already heard it).

A father and son are driving to the ball game when their car stalls on the tracks. In the distance, a train whistle blows a warning. Frantically the father tries to start the engine, but to no avail. The train hits the car, kills the father and injures the son, who is soon in an ambulance on the way to the hospital and the operating room.

As the boy is being prepared for surgery, the doctor walks in, expecting a routine case. However, on seeing the boy, the surgeon pales and mutters, "I can't operate on this boy. He's my son."

How could this be?

The answer to this story is of course, painfully obvious: The doctor is the boy's mother. It's right there in front of us, but we become wedded to one way of looking at a problem or to using one approach to define, describe, or solve it. This story and the **Hands Down** exercise help us to become aware of our own learning blocks.

Source: Original exercise inspired by Project Adventure

MENTAL MODELS

TEAM LEARNING

SYSTEMS THINKING

SHARED VISION

PERSONAL MASTERY

Web of Life

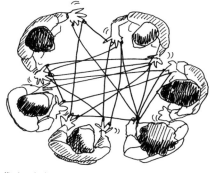

"We did not create the web of life,
we are merely strands in it."
Hekaka Sapa ("Black Elk")

"All persons are caught in an inescapable network of
mutuality, tied to a single garment of destiny. What
affects one directly, affects all indirectly."
Martin Luther King, Jr.

Web of Life is an eye-opening physical analogy for
the true complexity of many real-life systems. The
exercise shows clearly the myriad relationships within
particular complex organizational systems (such as a
school system, a marketing department, a research
lab). What is any group to do once it sees itself and its
environment in the light of such complexity?

In a fun, engaging way, the exercise provides a mirror
that reflects our automatic responses in dealing with
perplexing systemic issues. Do we respond with openness
and inquiry, or do we react with claims of certainty and
absolutes?

Discussions during and after this exercise take on a life of their own, often leading to a discussion of the inability of traditional management practices to respond effectively to complex, divergent problems. The **Web of Life** exercise helps people appreciate the benefits of systems thinking tools and concepts, and to see the need for the openness of a "beginner's mind" to think systemically.

⇨ To introduce a group to systems thinking concepts as a way to make sense of a group's complex interdependencies

⇨ To explore distinctions between convergent and divergent problems

⇨ To create an experience for participants to gain perspective on the complexity of their day-to-day interrelationships

☆ Appreciation for the potential of systems thinking tools to help us understand and navigate through "wicked messes" Connection between the need to be
☆ open (to acknowledge the "un-figure-outable-ness" of some complex problems) and systems thinking abilities Acknowledgment that some complex
☆ problems cannot be resolved with any single solution

Close the door for this one! You'll have a team mired in its own current reality, represented by a web of brightly-colored yarn draped in people's hands, maybe even around their waists and over their shoulders.

With a few carefully chosen questions, the **Web of Life** can be used to explore a variety of systems thinking-related topics. Debriefing this exercise includes a thorough discussion of three topics involving systems thinking. Try the exercise and see what systems related concepts and questions emerge.

Long-term teams using this exercise often "see" for the first time just how much of an interdependent whole they are and how many connections exist among themselves and within the larger system (such as their organization and their community).

To Run This Exercise

Number of People

8-12 works well. You can experiment with larger numbers

Time

Approximately 10-15 minutes, depending on the number of people

space

Enough space so your group can stand, nearly shoulder-to-shoulder, in a circle

Equipment

Ball of yarn (or string or surveyor's tape). Make sure it will unravel easily

set-up

Give someone in the group the large ball of colorful yarn

Instructions

1. 2. 3. 4.

Step 1: Ask group members to share the interrelationships that exist within their environment by referencing a specific problem, issue, or behavior they are all familiar with—perhaps the group's challenge is to improve fund raising efforts.

Step 2: They start by identifying key variables and assigning an individual to represent each variable. For example, Yolanda represents "time pressure," John represents "public relations," Nina represents "client's needs," Bob represents "regulatory pressure," and so on. (One person may represent two or more variables if necessary.)

Step 3: Now have one person suggest how her variable is related to one of the other variables, holding onto the yarn and passing the ball to the person who represents that variable. Have participants explain the

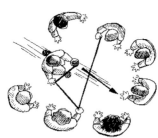

connection: Yolanda might say as she passes the ball of yarn to John, "When we have greater time pressure on the staff, there is less energy to invest in public relations, so public relations declines."

John takes hold of the yarn and pass the ball to Bob, explaining, "When public relations declines our image suffers, increasing public pressure for additional regulation."

Step 4: The group continues identifying as many connections as possible while the "web" grows in complexity. Once the group is sufficiently entwined in a "wicked mess" (ask, "Have you captured most of the important variables?) have them place the web, intact, on the floor where they are standing and return to their seats.

Debrief

With a few carefully chosen questions, this exercise becomes a terrific springboard into systems thinking-related topics such as:

- the distinction between convergent and divergent problems
- relationships and inter-dependencies, rather than simply objects and outside influences
- obstacles to learning in situations of complex interdependence

Convergent vs Divergent Problems

E.F. Schumacher, The British economist and author of *A Guide for the Perplexed*, makes a distinction between convergent and divergent problems. Convergent problems have a universally agreed upon solution: "the more intelligently you study them, the more answers converge." An example of a convergent problem is the need for a two-wheeled, human-powered means of transportation. "Various solutions are offered which gradually and increasingly converge until, finally, a design emerges which is 'the answer'—a bicycle—and an answer that turns out to be amazingly stable over time."

Divergent problems do not have "single" solutions. They are by nature situations with complex interdependencies. The more they are studied by intelligent and knowledgeable people, the more answers contradict one another.

A divergent problem example is the approach to educating our children. Advice from the most intelligent and experienced advisors varies widely, from "learn by doing is the best approach" to "passing on culture through experienced teachers is best." There is no one correct solution.

Some economists say that lay-offs are best for the economy, while others are convinced lay-offs hurt the economy. In the world of large organizations, divergent problems are often marked by high behavioral and dynamic complexity. Behavioral complexity is the extent to which there is diversity in the aspirations, mental models, and values of decision makers. In situations of high dynamic complexity, cause and effect are distant in time and space and causes of problems are not easily recognized through first-hand experience. Few players in the system have a sound understanding of the causes of the problem.

Through the **Web of Life** exercise, you can explore divergent and convergent problem solving from two perspectives:

- the nature of divergent problems

- our tendency to apply convergent problem solving approaches to divergent problems

In her book, *The Having of Wonderful Ideas*, educator Eleanor Duckworth points to the need for teachers to acknowledge and understand their students' thoughts and feelings when working with divergent problems. The art of facilitating divergent problem exploration is to keep the group from prematurely drawing a conclusion. The challenge, she says, is to "acknowledge complexity rather than replacing one simple way of doing things with another simple way of doing things—acknowledging the complexity and seeing where that leads." Listen carefully to the group's comments as the web becomes more intertwined and complicated. Remember or write down a few of their comments. When you refer to these comments, be careful to avoid identifying anyone, as who said what is not important.

After they've finished, ask the group for their reactions. How can they figure out all of the interactions represented by the web?

Note the different voices you hear. The authoritarian's voice will most likely emerge: "It's obvious, we have to do this and that...." Someone else will continue to suggest links that could be made between variables. Once you've heard most of the group, make the distinction between convergent and divergent problems. Point out the tendency to apply convergent problem solving skills to divergent problems. Share some of the comments made during the exercise, allowing the participants to see for themselves which comments reveal divergent or convergent problem solving approaches. Use this discussion to help the group step back for a different perspective and into a "beginner's mind;" to see the complexity of the situation without immediately trying to "solve" it.

Relationships and Interdependence

We've also used the **Web of Life** to explore and establish some of the key orientations of a "systems thinker." This exercise offers an experiential introduction to one of the primary orientations of a systems thinker: seeing relationships and inter-dependencies. Barry Richmond, founder of High Performance Systems, believes systems thinking skills are vital to our ability to solve today's problems. "If one accepts the argument that the primary source of growing intractability of our problems is a tightening of the links between the various physical and social subsystems that make up our reality, one will agree that systems dynamics and systems thinking hold great promise as approaches for augmenting our solution-generating capacity. The systems thinker's forte is recognizing interdependence."

Barriers to Learning

It becomes difficult to learn in situations of complex interdependence. If you've ever worked on a project team with internal and external team members located in diverse locations, communicating primarily by telephone, facsimile, and e-mail, you've experienced such a situation. It is particularly difficult to learn when any of the following conditions exist:

- significant delays between actions and the consequence of those actions

- multiple "feedback loops" (as opposed to simpler learning situations like balancing a check book, where a single feedback loop rapidly connects actions and observable consequences)

- significant "non-linearities" between actions and consequences (such as when small deviations from a norm produce no response but a slightly larger deviation produces a dramatic one)

resources

For more on how people learn in dynamic systems, see *Learning In and About Complex Systems* by John Sterman, an M.I.T. professor of systems dynamics.

To explore and understand these "barriers to learning," debrief the **Web of Life** exercise with one or all of these questions:

Where are there significant delays represented in the Web?

Where are there multiple feedback loops between variables?

Where might there be a significant disconnect between actions and consequences?

This set of questions, in addition to those you pose, may help the group to identify the organization's most chronic problems.

Source: Inspired by Outward Bound and "The Wall" exercise from *The Fifth Discipline* (pages 281-283).

Moon Ball

"All things are ready if our minds be so."

William Shakespeare

On the road to becoming learning organizations, we often encounter environments not typically conducive to learning. We may discuss the practical and important problems of learning in complex organizational settings while sitting in sterile conference rooms, dressed up in suits and seated around square tables. But what happens when you push that table to the side of the room, gather in a circle and pull out a brightly colored beach ball or two? When the beach ball comes out, faces light up, people remember fun times at the beach, and most find it difficult not to smile. That is a terrific mindset for entering an exercise geared toward learning.

In **Moon Ball** practically any size group can be challenged to keep the beach ball up in the air for as long as possible. Who would think such a fun and simple exercise could be such an effective practice field for learning and applying the disciplines of systems thinking, team learning, and mental models?

⇨ To provide an opportunity for participants to develop and test different problem-solving strategies. In the process they gain data on their ability to learn as a team

⇨ To put into practice the language of systems thinking by constructing a causal loop diagram of a firsthand experience

⇨ To experience the power of collective mental models

OUTCOMES

☆ Enhanced ability to apply the tools and concepts of systems thinking to real-life situations

☆ Improved fluency in the language of systems thinking

☆ An opportunity to practice team learning skills by exploring shared mental models in a safe, nonthreatening environment

CONTEXT

This exercise is particularly good for those with some exposure to the discipline of systems thinking. For example, suppose you and your management team have had an introduction to systems thinking through a two-day workshop, and you want to continue to build your systems thinking skills. Using the morning newspaper to practice drawing causal loops is one route. Playing **Moon Ball** is another. **Moon Ball** provides a commonly shared experience that lets a group reflect upon and understand different perspectives: from the event level to the patterns of behavior exhibited by the group, to the underlying systemic structures that drive the behavior and events.

Traditionally used as a team building exercise, **Moon Ball** is also ideal for considering the parallel processes of team problem solving and team learning.

resources

See **Community Maze** for a further explanation of levels of understanding (such as events, patterns of behavior, systemic structures).

To Run These Exercises

 Min: 8, Max: 30. ideal is 10 to 15. If you have more than 30 people, divide them into smaller groups, and use one ball per group

 12 to 20 minutes (depending on length of debrief)

 If you have access to the outdoors, use it. If not, you can run this exercise indoors, preferably in a large room with a high ceiling

 One beach ball, large balloon, or volleyball for each group. (A large balloon moves more slowly and is thus easier. Try to select the ball that will provide the right challenge for your group.) Flip chart and markers

 Clear away enough furniture so the participants can form circles, standing one to two feet apart and with four to six feet of space behind each person. Have the flip charts set up and the ball(s) nearby

Instructions

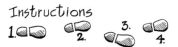

Step 1: Gather the group into a circle (or circles). Tell participants the goal: they must hit the ball as many times as possible while keeping it in the air for two minutes. Show how to hit the ball: move your arm in an upward arc and hit the ball with the palm of your hand, as in volleyball.

Step 2: Tell participants the details:

- They will have three tries

- In each, they will be given two minutes to plan and two minutes to implement their plan

- The score is the total number of "legal" hits during each try

Step 3: Tell participants the rules:

No person may hit the ball again until everyone else has hit it once

- They cannot hit the ball more than once per turn
  They can only use their hands to hit the ball

- Avoid heroic dives or other desperate efforts to keep the ball from hitting the floor

- The number of legal hits returns to zero if:
  —anyone hits the ball again before everyone in the group has hit it once during the turn
  —the ball touches the ground or other surface

Step 4: Tell the group to begin strategy planning. The facilitator keeps track of the 2 minute time limit.

If participants ask for clarification on the rules, it is important that you simply repeat the ones given in Step 3.

Step 5: Tell participants to put their plan in action! The facilitator keeps track of the 2 minute time limit. Each team appoints one of their members to count the number of times the ball is legally hit.

Step 6: After 2 minutes tell the participants to stop. Ask them their score, then begin the second try, starting with step four.

 If the teams repeat the exercise enough times, you will have a group who devises a plan to pass the ball around the circle, hand to hand. This is not allowed.

Variation

If you are conducting a longer workshop, you can do one **Moon Ball** try a day, rather than all three at once. This permits more reflection on ways to improve. It also gives the team an opportunity to test and observe the influence of various skills and attitudes they are learning during the workshop.

Debrief

The group usually improves with each attempt, as participants learn to work together. At first the group may approach the exercise randomly, playing the game as a group might play volleyball—they hit the ball up into the air and hope there will be someone there to hit it as it comes down. Each iteration of planning and execution should show improvement in strategy.

On the second try, the group may designate a hitting order (who hits the ball to whom). Often this will improve their score. On the last attempt, the group may realize that if they stand near the person to whom they are hitting the ball, the score will be further improved. Hopefully, there will also be improvement as the individual members of the group learn how to keep the ball under control while hitting it. The person running the exercise (whether he is a facilitator, manager, or colleague) should listen closely for comments and phrases from the group as they move through "solving" this challenge. Write down what you hear and incorporate those phrases and comments in the post-exercise discussion.

We've used the following four-step process to effectively debrief **Moon Ball** (see Intro for more detail).

Step 1: *Tell the Story* Ask the group to describe what happened. If you were running two groups simultaneously, you may want to bring them together for a large group discussion. Write some of the key phrases on the flip chart. You may also want to review your notes, and include some of the comments you overheard.

Step 2: *Graph the Variables*
What was the behavior they observed over the three trials? Ask the group to plot their data for trial #1, #2, and #3. In general, the graph of the variables (the reference mode) will look like the one at right.

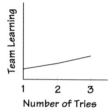

Once the group has plotted the behavior mode, it's quite useful to show three different reference

---

modes—learning, stagnation, and worsening performance—as a comparison.

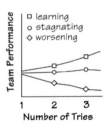

Step 3: *Draw the Causal Loop Diagram* The nature of a group's progress is not always linear; it is not guaranteed that it will consistently improve or decline over time. As with any kind of learning, the group may have a burst of insight and leap forward, or it may hit a plateau.

The next logical question is: "What causal loop structure included in the relationships could explain the differences among the three modes of behavior?"

Provide the group with a causal loop diagram. It contains a fairly general set of terms that can be used to describe most groups' success or failure in trying to improve their performance.

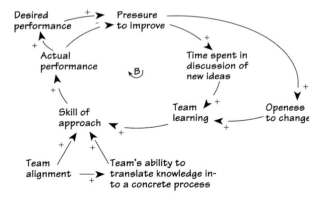

 If your participants have practice with creating causal loop diagrams, you may want to challenge them to first create their own causal loop diagram and then compare and contrast their diagrams with the one above.

Step 4: *Identify the Lessons*
Once you and the group members have discussed each of the factors and relationships in the causal loop diagram, discuss how these manifested themselves during the group's efforts to learn how to succeed in **Moon**

Ball. Then you may want to ask the group what they considered to be the constraints in this activity.

What is the limiting action? Ask this question and you'll receive a variety of answers. One limiting action here can be that participants tenaciously pursued the same approach, without stopping to reflect on their assumptions, hear other ideas, or consider other options.

Another common problem is that the group members fail to develop a plan that takes into account the physical properties of the ball. Another may be coordination: the group fractionates into several subgroups, each making its plan without involving the others.

Through this discussion the group often realizes that their own interpretations placed self-imposed limits on their ability to solve the problem. By focusing on their thinking processes, they might find leverage points (an area in a system where a relatively small change can produce noticeable improvements in a system's behavior).

What small change in the way the group implements their strategy could improve their results?

In **Moon Ball**, one limiting action and possible leverage point lies in the group's mental models about how to keep the ball up in the air. For many, past experience with volleyball games influences their approach to this new challenge: it resembles volleyball, yet the goal is different. A solution requires the group to think outside the "volleyball game" box. For example, in volleyball, players tend to stand rather far apart and hit the ball high to set it up for a "spike" by a teammate. Both the distance and the high-hitting are counter-productive in **Moon Ball**.

Barriers to Learning

Did you hear comments discouraging any solution that strayed from the volleyball approach? You can serve the group best by feeding back to them those ways in which they block their own learning. This is particularly useful for long-term teams who see themselves becoming a learning team. One question to ask is "What are some blocks to surfacing and changing our mental models?"

Here are a few potential blocks:

- We apply our mental models unconsciously

- We don't want to risk "losing" or being "wrong"

- We don't want to risk upsetting or embarrassing others

- When confronted with results that don't match our desired outcomes, we look for information that confirms our ineffective strategy. As a result, when we make errors we are unlikely to discover them.

Were any of these blocks apparent during **Moon Ball**? How can we see these blocks operating in every day life?

VOICES FROM THE FIELD

Don Seville, a system dynamicist with an affinity for environmental issues, tells us that **Moon Ball** is an "excellent real time practice in the movement between events, patterns, and systems structures...participants are only initially aware of the mayhem of reacting to the ball. And then, after they've tried the exercise several times, patterns of behavior begin to emerge. When the group takes the time to slow down and look at the patterns, through drawing behavior over time graphs, they begin to see how their own behavior is part of the system, and begin to explore how they might intervene in the system. The participants can easily see how their behavior is part of the system and they associate breaking links with things that they themselves do."

Source: Fred Kofman, founder of Leading Learning Communities, originally identified several of the learning barriers described in this exercise.

Paper Tear

"Remember playing 'telephone' and being delighted and amazed at how the message got distorted with only a few players? At a young age, we knew information for its dynamic qualities, for its constantly changing aliveness. But when we entered organizational life, we left that perspective behind."

Margaret Wheatley

When stake holders come together in a group, they frequently discover they have different perceptions of what's going on. This is especially true when the group comes together to understand and intervene in a complex system. As divergent perspectives become evident, there is a tendency to increase the frequency or the volume with which we express our own thoughts. We should try to empathize with the listener and discern what they are understanding and what might be the sources of misunderstanding. We assume that if we do an eloquent job of describing our thoughts, the hearers will end up with the same images in their minds. This simple exercise shows quickly how unfounded this assumption really is, even when the hearer shares your goals and has a strong incentive to understand your meaning.

⇨ To make the distinction between hearing (the biological process of assimilating sound waves) and listening (adding our interpretations of what is being said)

⇨ To demonstrate the importance of effective communication skills and listening skills to thinking and acting systemically

☆ Heightened listening and communication skills, which will help participants to create causal loop diagrams that are less abstract and more accurate

☆ Increased awareness of the multiple interpretations that can be obtained from the same message

Your executive team gathers for an all-day meeting to conceptualize a systems view of a company-wide problem. When asked to describe the problem's contributing factors, one capable, respected and energetic team member says, "I have a list as long as my arm!" and proceeds to identify the key variables she believes are critical to solving the problem. Another team member jumps in, adding his thoughts to the list, while refuting several of the suggestions made by the first member. Before the first speaker can respond, another person has added his thoughts.

This scenario is common, especially when a team, with multiple perspectives, is trying to build a shared picture of a complex system. **Paper Tear** can act as an excellent and brief warm-up for such a team. It brings the group's attention to two critical skills that will enhance their abilities to solve problems systemically:

- communicating to be understood, and

- listening to understand.

One area of great leverage is the way we conduct meetings. A research study from Yale University showed that 75% of comments made in average business meetings are statements, opinions and criticisms.

When acting in this way, a team blocks its own ability to think and act systemically. The effects of such a group problem-solving process can be remarkably damaging. Dialogue and skillful conversation are forms of conversing in which participants seek to balance advocacy and inquiry, suspend assumptions, and build a collective understanding. By providing a mirror to the fragmented nature of thought processes, dialogue can help bring a more holistic, systemic perspective of the issues at hand.

To Run These Exercises

Min: 10, Max: 50, Ideal: 10 to 20

5 to 10 minutes (depending on length of debrief)

None

3 sheets of 8.5 x 11 inch paper for each person

Pass the pile of paper around the room, and ask each participant to take three sheets. Keep one sheet for yourself

Instructions

Step 1: Don't start the exercise until everyone has a sheet of paper in their hands. Have participants sit someplace in the room where they can hear you.

Step 2: Tell participants the rules: There is no talking. Participants must close their eyes and do exactly what the facilitator says to do. State that the goal is for everyone to produce identical patterns with their pieces of paper.

Step 3: The facilitator reads the following instructions, slowly and distinctly: "Fold your paper in half and tear off the bottom right corner of the paper." (Pause and allow the group to try this.) "Fold the paper in half again and tear off the upper right hand corner." (Pause) "Fold the paper in half again and tear off the lower left hand corner." (Pause) "Open your eyes, unfold your paper, and hold it out for the group to see."

Step 4: Participants look at what they have produced and what others have produced in comparison.

Step 5: Ask the participants to form a group of three. Have them repeat the exercise as a trio, with these instructions:

- Tell them to pick two people to be the listeners (with eyes closed or backs turned toward the communicator) and the other person to be the communicator. Ask them to repeat the exercise with the communicator giving the listeners instructions on how to fold and tear the paper. The communicator does not have to give exactly the same instructions you gave the first time. But he does have to go through at least three steps, each involving a fold and a tear. After they've completed the exercise, ask the listeners to compare their sheets of paper. Are they similar? The majority will be different.

- Ask them to switch roles, with the communicator becoming a listener, and a listener becoming a communicator. This time, allow the listeners to talk. Then proceed with the directions in the paragraph above.

After they've finished, ask them to discuss what they may have noticed about their listening and communication abilities. What did they notice in the three attempts? Did they become more accurate? If so, why?

Debrief

Most often, each participant creates one of four or five different shapes out of their papers, some resembling the paper snowflakes that children make during the winter. Participants are likely to be surprised by the different interpretations of the same, simple message.

Ask participants to describe their experience. How would they describe their listening skills? How effectively did they communicate? What would have happened if they asked more questions before the exercise began? What would be an analogy of this experience in a real organization?

Ask the group to consider where they experienced or saw less dramatic but similar examples of errors in communication that led to unwanted results. What is their work environment like? In general, how are questions perceived? Are they encouraged or discouraged?

The important point here is that even at times when we think we are listening or communicating clearly, errors may still occur due to misinterpreting what someone else has said or someone misinterpreting what we say. By improving our communication and listening skills, we improve our ability to think systemically and, in group settings particularly, to learn as a team.

VOICES FROM THE FIELD

Dennis Meadows has used **Paper Tear** numerous times and offers this variation on the debrief: "I ask why the group got such diverse results even though everyone was honestly trying to follow instructions. Usually one participant will have views on this, sometimes even blaming me for the ambiguities. Then I say, 'Okay, you demonstrate, but with the same rules — no talking and everyone has their eyes closed. You also have to go through at least three steps each involving a fold and a tear. Go ahead and show how you would get everyone to produce the same shape.'

Maybe he or she will succeed; typically they don't. But the very elaborate and detailed communication they attempt to achieve this simple task illustrates, whether they fail or not, just how hard it is to make one-way communication effective in conveying mental images."

Belief Release

"When we come to a point of rest in our own being, we encounter a world where all things are at rest, and then a tree becomes a mystery, a cloud becomes a revelation, and each person we meet a cosmos whose riches we can only glimpse."

Dag Hammarskjold

Robert Hanig, a master in the field of organizational learning and dialogue, adapted this exercise from the work of Robert Fritz (author of *The Path of Least Resistance*) to help dialogue participants become more "in the present." We have found this to be an extremely valuable exercise when "reframed" for a group preparing to use systems thinking concepts and tools. When it is carried out in a systems thinking context, it helps us to see our propensities to be preoccupied with a single issue or problem or factor and thus fail to see the whole system.

⇨ To provide an opportunity for individuals to become more "in the present"—to put the "to do" lists of the day on the back burner and to focus on surfacing and aligning with personal visions

⇨ To combat mindless, automatic categorizations and judgments

⇨ To promote mindful creation of new distinctions and categories

☆ increased clarity of thought

☆ An enhanced awareness of and attention to habitual processes of mental model construction

A group sits around a flip chart covered with partially constructed causal loop diagrams. You look around the room and see perspiration on a few brows, pupils larger than normal and shoulders gathered up near the ears. Statements of advocacy reign, and inquiry is nowhere in sight. The group is trying to understand the complex set of inter-dependencies between their business unit and the parent company, and as one person says in exasperation, "Trying to make sense of all these different opinions is like trying to herd cats!"

This may be a good time for **Belief Release**. The exercise helps groups get "unstuck" by encouraging them to slow down, step back, and notice their preconceived assumptions and solutions. As ossified beliefs are suspended, intuitive abilities are awakened and encouraged. Robert Hanig explains, "The process of noticing and letting go of preconceived notions of a particular system provides more than just mental clarity. It provides the physical, emotional and spiritual clarity where new dimensions are allowed to emerge."

To Run These Exercises

Any number

Approximately 10 to 15 minutes

space Enough so that people can sit comfortably in chairs or on the floor

 Equipment A good imagination and a script similar to the one that follows. Once you get the hang of it, you will want to adapt the sequence of visualizations to suit the situation and the group

 set-up Arrange chairs in a circle, clearing away tables and other furniture. To further "ground" the exercise, we suggest that participants uncross their legs, plant their feet firmly on the ground, and put on the floor anything they might be holding

Instructions

Prior to conducting this exercise, walk through it yourself. Do whatever you need to do to bring a spirit of open-mindedness and presence to the group. The more present you are, the more engaging the exercise will be for the participants.

After you have prepared yourself, sit with the group and ask them to do the following:

"Hold onto your chair, hold it as tight as you can, imagine that if you let it go, you would float up and hit your head hard on the ceiling. When I ask you to focus on a belief or assumption and hold it, hold it the way you just held your chair, as if your very existence depended on it.

Now release your chair, letting the muscles relax in your fingers, your arms and your shoulders. When I ask you to release or unhand the belief or assumption, release or unhand it the way you released your chair, just let it go, let it float away.

Now, relax, close your eyes and take a deep breath. In the privacy of your own mind, begin to focus on a negative belief you have about yourself or a weak point, where you are not as good as you could or want to be...really hold it tightly, as if your very existence or identity depended on it, hold it (7 seconds)...now release it...un-hand it...just let it go. Imagine that it moves away from you and goes where it will.

And now focus on another negative belief or opinion you have about yourself, or something you are ashamed of... really hold it tightly, as if your very existence depended on it, hold it (7 seconds)...now release it... unhand it... just let it go.

And now focus on a positive belief or opinion you have about yourself, or something you like about yourself, something you are proud of. Really hold it tightly, as if your identity depended on it, hold it (7 seconds)...now release it...unhand it...let it go.

And now focus on another positive belief or opinion you have about yourself or something you like about yourself, something you are proud of. Really hold it tightly, as if your identity depended on it, hold it (7 seconds). Now release it...unhand it...let it go.

And now focus on a negative belief or opinion you have about other people, something you don't like or find worthwhile...really hold it tightly, hold it...now release it... unhand it...just let it go.

And now focus on a positive belief you have about other people, something you like or admire...really hold it tightly, hold it...now release it...unhand it...just let it go.

And now focus on a negative belief you have about the way the world is or about the way things are, something you think is a major and perhaps unsolvable problem... really hold it tightly, hold it...now release it... unhand it... just let it go.

And now focus on a positive belief you have about the way the world is or about the way things are, something you think is wonderful and gives you hope for the future... really hold it tightly, hold it...now release it... unhand it... let it go.

And now focus on a belief or opinion you have about

another member of this group...really hold it tightly, hold it...now release it...unhand it...just let it go.

And now focus on a belief or opinion you have about one of the facilitators....really hold it tightly, hold it...now release it...unhand it...just let it go.

And now focus on the experience you are having right now...really hold it tightly, hold it...now release it...unhand it...just let it go.

And just begin to become aware of your feeling, thoughts, and experiences...hold them for a moment, notice their characteristics, notice their origin...and just let them go...just move along.

Take another breath, relax your focus and open your eyes."

Debrief

It is difficult to write a linear debrief for this exercise; there are so many directions a group may want or need to go after experiencing it. If they are moving into "problem solving mode" (either to brainstorming on the critical variables of their particular systemic issue or back to exploring the dilemma through causal loop diagrams), encourage them to maintain the sense of awareness and clarity they may have gained from the exercise.

Belief Release also creates an opening for a discussion about the importance of intuition in systems thinking. Consider the novice rock climber who at first uses upper body strength and stubborn determination to muscle up a steep rock face. Eventually the climber wears out and discovers that balance, grace and pacing are equally—if not more—vital skills for rock climbing.

There is a similar analogy in systems thinking. "It's not just about reasoning your way around the loop," explains Robert Hanig, "It's standing back and intuiting where the real leverage is...it is being able to stand back from an organization and just be present with it. If you do this, you notice your automatic solutions and decide if they will lead you to high-leverage interventions."

In his book, The Heart Aroused, the poet David Whyte shares a remarkable Native American Indian axiom. It was given by an elder to a child whose life depended on the answer to his question, "What do I do when I'm lost in the forest?"

LOST

Stand still. The trees ahead and bushes beside you
Are not lost. Wherever you are is called Here,
And you must treat it as a powerful stranger,
Must ask permission to know it and be known.
The forest breathes. Listen. It answers,
I have made this place around you,
If you leave it you may come back again, saying Here
No two trees are the same to Raven.
No two branches are the same to Wren.
If what a tree or a bush does is lost on you,
You are surely lost. Stand still. The forest knows
Where you are. You must let it find you.

The wisdom here is that living with complexity has everything to do with paying attention, with "looking the world right in the eye even as we are creating our products, services, gifts," says David Whyte. **Belief Release** can help us to maintain perspective while recognizing our connections to the whole.

 resources *The Heart Aroused,* by David Whyte, (page 259-260) To obtain a copy contact Currency/DoubleDay at (800) 223-6834.

After **Belief Release** you may wish to explore how the members of the group are actually a part of the system they are trying to "fix." "You are a part of the system," Robert Hanig says. "The more you try to look from the outside in, the more you may miss the connections between you and the system. When you actually stop and feel the connections in you as well, it becomes a part of you, you have a much more intimate and holistic perception, you can feel your way into it." As the Indian philosopher Krishnamurti reminds us, "The observer and the observed become one."

If this exercise intrigues you, we urge you to read Fritz's *The Path of Least Resistance* prior to using the exercise for the first time. You may also want to explore Fritz's Technology for Creating® curriculum. Call (800) 848-7176.

Source: adapted from the work of Robert Fritz

Balancing Tubes

"The present, therefore, has several dimensions. . . the present of things past, the present of things present, and the present of things future." St. Augustine

One of the most important foundations for any study of systems behavior is the concept of time horizon. This is the interval of time over which the system manifests the full pattern of behavior that concerns us. Or, described in another way, it is the length of time required for us to see the system's full response to any action we might take. A change in corporate capital acquisition strategy may have a time horizon that spans several years, whereas the time horizon for a change in an advertising campaign may be a few weeks.

It is very important for the members of a group to focus explicitly and agree on the time horizon that should characterize a specific study. Miscommunication, misunderstanding, and even conflict among group members arise when they've all implicitly adopted different time horizons for an issue.

Challenging mindsets about time is, without a doubt, more than an intellectual exercise. The **Balancing Tubes** exercise provides an exquisitely simple, physical experience from which participants increase their awareness and understanding of appropriate time horizons.

⇨ To introduce the concept of time horizons

⇨ To illustrate that control comes only when an appropriate time horizon is determined

☆ Increased familiarity with the language of systems thinking

☆ Opportunity to consider time horizons in a real-life situation

☆ Opportunity to practice drawing behavior over time graphs

It is crucial to focus people's attention very early on the concept of time horizon. You can do this by using the **Balancing Tubes** exercise and then following with a comparison of different time horizons for a specific product (see the following debrief on copper commodities as an example).

This exercise in itself is simple and therefore cannot convey all the important aspects of the Time Horizon concept. But it does make this important point: when you are trying to understand and control a dynamic system, there will be an appropriate time horizon within which your observations can lead to insight and to effective management. Focusing on changes in the system that occur over shorter or longer periods than the appropriate time horizon will not give you control.

To Run This Exercise

Any number

5 minutes for the exercise itself, the discussion after the exercise may be 10 minutes or longer

---

space 3 - 4 feet between each person. It's better to have the group standing in a circle, so each person can see the performance of the other

 Equipment One paper tube for each participant. We have also seen groups using 3 foot long sticks or cardboard tubes

set-up Prepare enough tubes ahead of time. Using newspaper or newsprint, begin at one corner and roll diagonally around a broom handle. Slide it off and tape it, so you have a 1-inch tube about 3 feet long.

You might place one tube at each seat prior to the participants entering the room. Or keep them in a paper grocery bag and quickly pass them around to the participants just before beginning the exercise.

Instructions

1. 2. 3. 4.

Step 1: Tell participants, "Your goal is to balance this tube on your fingertips." Demonstrate balancing the tube vertically on your fingers, palm up. First balance the tube while focusing your eyes on a spot just 1-inch above the point where the tube meets your fingers. Pause to give the group time to try this.

Step 2: Now, balance the tube while focusing your eyes on a point at the top of the tube. Pause while the group tries this.

Step 3: Finally, try to balance the tube while focusing your eyes on the ceiling. Wait for your group to try.

Participants will find it difficult or impossible to balance the tube of newspaper on their fingers when looking at a spot that is either too close to their fingers or too far away.

Debrief

Here are a few questions to begin with:

Which of the three methods worked best?

Why do you think it was easiest to balance the tube when focusing on the top of the tube?

What was changing when you shifted your perspective?

The main factor that changes when you shift from one focal point to another is the length of time between when the tube starts to fall off balance and when your eye detects the movement and provides the information required for your hand to adjust. This is because the tube must move a certain distance before your eye can detect that it has changed position (This is sometimes referred to in psychological experiments as the JND—the Just Noticeable Difference.) When you focus on the bottom of the tube, the top must move a great distance to provide the JND stimulus you need to counteract the fall. Typically, you will be too late, and the tube will fall. When you focus on the top of the tube, the top needs to move only a little to give you the JND movement you need to cope. So your response is relatively quick and usually effective in maintaining balance. Of course when you focus on the ceiling, the tube can fall almost completely off your finger before you will notice its movement, and there is practically no control at all.

Balancing Tubes makes two points: 1) the appropriate time horizon is dependent on your reason for tracking a particular system and 2) if our time perspective is too short or too long, we won't be able to control satisfactorily the behavior of the system. One example of this is provided by movement in commodity prices. If you study the movements in the price of copper futures on the commodity exchange you can see three quite distinctive patterns of behavior, each with an extremely different time horizon.

The first reference mode is hourly and daily fluctuations. It looks a lot like noise, and the changes are caused by

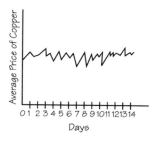

Days

adverse weather that interferes with mining, by strikes that may reduce the labor force, by interruptions in transport that may cause temporary shortages in important regional markets. If you were to record the average hourly price of copper over time, you would get a reference mode that looks like the diagram above. It has a time horizon measured in days or weeks.

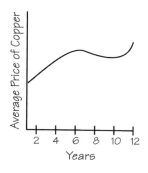

Years

The second reference mode is the seven year production cycle. It looks like a sinusoidal wave, and it is caused by producers' and consumers' delayed responses to price changes. If you were to record average monthly or quarterly copper prices over time you would get a reference mode that looks like this diagram. It has a time horizon of 5 to 10 years.

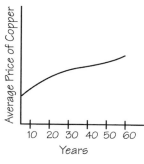

Years

The third reference modeis the long-term growth or decline in the relative price of copper. It looks more- or-less like a straight line, and is caused by the rising costs of energy required to produce the material, by depletion of low cost ores, and by the impact of technological advances that change the amount of copper used in different products. If you were to calculate the average price of copper over the past 7 years and record that number each year, you would get a reference mode that looks like the diagram here. It has a time horizon of 20 to 50 years or more. All three reference modes plot the same variable: price of copper. But the differences in the time horizons make the three diagrams of interest to very different clients. Speculators and manufacturers who use copper are concerned with the first reference mode. Those selling mines or mining equipment, or contemplating the construction of a new factory that will rely heavily on copper as a raw material, care about the second. The third would possibly be of interest to those in senior government positions within nations that depend heavily on copper for their export earnings.

You could imagine three people sitting down and agreeing that they need a better understanding of what causes change in the price of copper. But each has a different time horizon, making it impossible for them to agree on how to proceed. Policy options that are very attractive to one of them would seem irrelevant to the others.

VOICES FROM THE FIELD

A great observation comes from a group of 11 year old students in the Catalina Foothills School District in Arizona. One of the school's teachers decided to use **Balancing Tubes** to explore the dynamics of the Civil War (which the class had just studied). Without explicitly mentioning the concept of "time horizons," the students discussed how General Scott was the only one who thought the war would last a while. As a result he was dismissed. The kids said that all the others were "looking at the hand" while General Scott was looking at the "whole picture" and "looking ahead."

Source: Original inspiration for **Balancing Tubes** provided by Benny Reehl and John Shibley

Touch Base

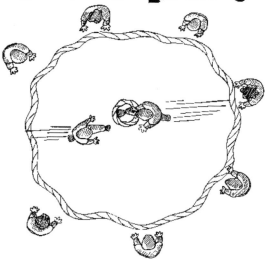

"Dynamic thinking is the ability to see and deduce behavior patterns rather than focusing on, and seeking to predict, events. It's thinking about phenomena as resulting from ongoing circular processes unfolding through time rather than as belonging to a set of factors."

Barry Richmond

When systems dynamics students are challenged by their teachers to think "dynamically," they are often asked to think about everyday events or newspaper stories in terms of graphs plotted over time. **Touch Base,** a get-up-and-move kind of exercise, is good for helping people think dynamically. It is also a challenge that tests and strengthens teamwork and communication. There is a race against time and a very small space through which a whole team must pass without touching one another. This exercise may be repeated several times in succession, like **Moon Ball**, to test and illustrate the capacity for learning. Or you can devote more time to the initial planning phase and just use it as a one-time problem solving initiative.

⇨ To experience the connection between improved communication, team learning, and systems thinking

⇨ To show that the path to improvement may require stepping back to get a different perspective on the problem and the solution

☆ Understanding of the elements of an effective team learning cycle

☆ Practice being a part of the system and observing the system

☆ Enhanced team spirit and fun

Without fail, this exercise starts off slowly as the group takes its task seriously: getting each person to touch the center of the rope while crossing to the other side of the circle, without touching each other. Within minutes after the first test run, the laughter starts and the ideas flow and by the end, we wish we had a bull horn to cut through the din to get everyone's attention. **Touch Base** is one of the more physically engaging exercises in this volume. Therefore it is a terrific way to give a jump start to the morning or to shake out post-lunch fatigue.

To Run This Exercise

 Min: 20, Max: 50, Ideal: 20 to 25

 15 to 30 minutes (depending on length of debrief)

 Conduct this outdoors, if possible. If not, clear furniture to the side of a large room

 60 to 90 feet of rope (enough to make a circle 20 or 30 feet across)

Create a circle on the floor with the rope, large enough for the participants to stand around with 2 to 3 feet between them. Place a Frisbee or paper plate in the center.

Instructions

1.   2.   3.   4.

Step 1: Gather the group in a circle around the large rope circle. If you have more than 35 people, divide the group in half and set up two circles of rope.

Step 2: The goal is for the whole group to complete the exercise as quickly as possible. Participants must leave their positions outside the large circle, touch the disk in the center, and arrive outside the large circle at a position opposite from the starting position. If any two people touch each other, both must go back to their original positions and start again.

Step 3: Tell participants that the clock starts when the first person crosses into the large circle, and it stops when the last person gets back out to a place opposite where they started.

Step 4: Allow the group 5 minutes to plan and 10 minutes to execute the plan. Or if you wish to use it as an illustration of learning, run the exercise three times, giving 5 minutes each time for planning and 3 minutes to run the exercise. Keep track of their shortest time to complete the process during each three-minute trial. Plot those results on a poster sheet or flip chart.

 NOTE As you should with any activity, keep an alert eye on safety. When this activity speeds up, there is the potential for tripping and bumping into one another. If necessary, stop the activity in mid-stream and announce time penalties for anyone who runs.

Debrief

There are several different strategies for accomplishing the goal quickly. But for each strategy, the principal insight needed to improve performance is to recognize that someone should be assigned to stand back from the circle and observe the overall pattern of traffic flow. When the entire group is engrossed in the task of getting from one side of the circle to the other, without touching each other, it is hard to see what factors are slowing the group as a whole.

Touch Base is well-suited to a debrief based on the four-step debriefing process described in the introduction. After the group tells the story of what happened and identifies some of the key variables of their problem-solving process (Step 1), you can take them through a graph of those variables and a causal loop diagram (Steps 2 and 3. See the **Moon Ball** exercise for more details.). Finally, help them look for insights (Step 4).

Here are a few questions to ask:

> What was the group's strategy? Did they work as a team or did individuals act independently, leading to a confusing mass of people?

> Was the group careful to make sure everyone understood the plan before starting?

> Did everyone feel that they had a chance to express their own ideas and suggestions?

> What other aspects of teamwork were left out?

> Would the group say that it learned over the consecutive trials?

Did the group assign anyone to step back and view the process as an observer? This other perspective could have enhanced the team's ability to learn from each trial.

Stepping back to look at the whole is one of the characteristics of those who think systemically. Thinking systemically means that you look at everything that is happening, not just what is happening from an individual, isolated viewpoint. And a systems thinker consciously experiments with different time and spatial perspectives to determine which ones will give the most useful insights into the problem. One of the most important connections to make to systems thinking, is to link the learning and improvements the group members may have experienced to their abilities to observe behavior patterns rather than events.

Ask the group to think of examples from work where looking at the event actually caused them to complicate the problem. How could they have looked at the whole system instead?

Squaring ·the· Circle

"New organs of perception come into being as a result of necessity. Therefore, O, man, increase your necessity so that you may increase your perception."

Rumi

During **Squaring The Circle** a team engages in a process that may feel a lot like real life—trying to develop a totally shared view of their problem, in the dark. In this exercise, the group is blindfolded (or told to close their eyes) then asked to solve a problem that can only be solved with the assistance of all team members. Success hinges on members of the group calling on their visualization skills and understanding their contribution to the solution, both individually and collectively.

In nature and society, successful systems are those that possess a self-organizing capability—those that have capabilities to act autonomously, to view themselves in relation to their environment, and to adapt accordingly. **Squaring The Circle** challenges a group to become its own self-organizing unit, and to find its own order (which is achieved when they "square the circle") through team work, shared visualization, and systemic thinking.

⇨ To explore experientially the meaning of team learning

⇨ To introduce the concept of self-organizing teams and systems

⇨ To examine what occurs when communication is limited to the voice

☆ An appreciation of the process necessary to create a share vision

☆ An appreciation for how an understanding of the whole improves team work and problem solving

Traditionally used a team building exercise, **Squaring The Circle** has all the makings for a practice field rich in opportunities to explore the concepts of systems thinking, team learning, and self-organizing teams. Deprived of sight and thus of a wide array of non-verbal communications (for example gestures and facial expressions), the team is challenged to adapt to their new environment (a challenge faced by all self-organizing teams) and through the skills of team learning, create new methods of communication and problem solving.

To Run This Exercise

Min: 8, Max: 30

20 to 30 minutes (depending on group)

Outdoors or in a room large enough for participants to form a loose circle

One long rope, 10 yards or longer

Have the rope nearby and make sure it can be easily uncoiled without having to unravel tangles and snarls. Ideally it should already be uncoiled and ready on the floor

Instructions

Step 1: Have everyone line up, shoulder-to-shoulder, in a straight line, all facing in the same direction. Ask the participants to put their hands out in front of them, palms up. Place one end of the rope in the hands of a person at the end of the line and walk down the line having each person take hold of the rope with both hands. At the end of the line, turn around and walk back up to the original end, but this time just playing out the rope on the floor. Then tie the two ends of the rope together. Now all people are bunched on half of the loop.

Step 2: Tell participants the rules: "The entire rope needs to be used. Close your eyes, and keep them closed during the rest of the task. You may slide along the rope, but you cannot change positions with anyone else on the rope. When you personally think that the group has finished its task, raise your hand and I will ask for a vote. If a majority of the group thinks you are finished, I will ask you to stop and open your eyes. Otherwise, I will tell you to keep going."

 NOTE If a participant doesn't want to close his eyes, or accidentally opens them during the exercise, ask him to let loose of the rope and step back silently. He will serve as an observer who can later help the group understand the strengths and weaknesses of their problem-solving approach. You can also ask one or two people to volunteer to act as observers, prior to the start of the exercise.

Step 3: Finally say, "Your goal is to create a square while everyone maintains their hold on the rope."

Step 4: As the facilitator, you should see that no member of the group wanders into anything (a wall, a tree or a hole). As the group attempts to solve the problem, you should remain silent. When a participant raises a hand to signal the process is complete, the facilitator asks the group to vote on whether they are finished. If the majority believes the task is accomplished, ask them to open their eyes. Have them place the rope on the ground, being careful to maintain the shape.

Step 5: Give the group a chance to look at the shape of the rope and then move to a comfortable place to sit and debrief. Leave the rope on the ground so that the group can refer to it during the debrief.

Variations

If you have a small group, about 6 to 10 people, you may opt for "Triangling the Circle." Instead of making a square out of the rope, participants must create an equal-sided triangle.

Debrief

Some groups create a perfect square, some a triangle and others a shape that looks like an amoeba. Whatever the shape, you and the group can be assured that there is learning to be had. If you had any observers, allow them to comment on what they observed. Then ask participants to describe their experience:

How effective was the group's communication?

What was their strategy?

Was the strategy effectively communicated?

Their strategies will vary. Some group members will figure out that they can make the process easier, if they count off and try to align the group so that there is an equal number of participants on each side of the square (all sides are equal). Other groups will figure out that the process can be improved if the people who make up the corners are chosen.

Team Learning

Squaring the Circle provides a good opportunity to explore how the group may have learned over the duration of the exercise.

Revisit what happened in the first few minutes of the activity: How does this compare to what was happening toward the end? How did the group improve? Ask participants to use the elements and the links in The Learning Cycle (see diagram) to explain their success or failure in turning the circle of rope into a square.

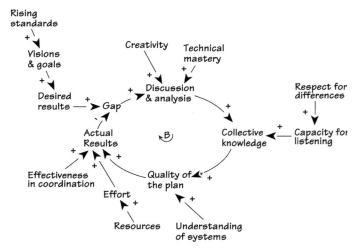

Self-Organizing Teams

To explore the concept of self-organizing teams, you might consider these questions:

Did a leader emerge?

How did the leader or lack thereof affect the group dynamics?

How did "not being able to see" affect your ability to communicate?

We've often witnessed this somewhat ironic occurrence: after moving almost immediately an effortlessly into a good square, members of the group start to analyze and intellectualize the process ("It seems too easy," one person said in a post-exercise discussion) and their satisfactory solution deteriorates. In the end, the square becomes more misshapen than the group's original solution.

If you are working with a long-term team, you may also wish to explore the analogy to communicating without being able to see one another. How are the dynamics similar for a team, such as a far-flung sales group, whose members are constantly on the road and who communicate mostly by telephone, fax and e-mail?

resources For more on self-organizing systems, see "Change, Stability and Renewal: The Paradoxes of Self-Organizing Systems," *Leadership and the New Science*, by Margaret Wheatley. To obtain a copy contact Berrett-Koehler Publishers, Inc. (800) 929-2929.

Systems Thinking

From the perspective of systems thinking, **Squaring The Circle** is like redesigning a system: the process will be more effective if individuals in the group listen to everyone and ensure that all understand and agree with the plan before taking action. This exercise shows how futile it is for an authoritarian "leader" to try and dictate to everyone else what they must do to create a square.

1·2·3 GO!

"Experience is structured in advance by stereotypes and idealizations, blurred by caricatures and diagrams."
Mary Catherine Bateson

This exercise is not meant to promote deep learning as much as it is to provide a wake-up call to the brain. After an experience with **1-2-3 GO!** the message your brain hears might sound something like this: "Stay alert! (Am I operating on automatic pilot? Have I made an assumption?)" The exercise is so simple to facilitate, yet so effective. In many ways, it supports the aphorism many have heard since childhood: actions speak louder than words. We have an affinity for this exercise, because it reaches the tacit parts of our brains that traditional learning methods find difficult to touch.

⇨ To illustrate the speed with which people often move from forming assumptions to taking action

⇨ To emphasize the importance of listening before acting when working in a team

⇨ To show that everyone needs to examine the congruence of their actions with the message they are trying to convey

☆ Increased awareness of the need to introduce a "space" between thought and action

☆ Appreciation that behavior in social systems depends on more than what is spoken and written (when developing a system model don't just listen to what people say, watch what they do)

1-2-3 GO! pulls its weight in the "warm up/awareness generating" category of exercises. During a day or afternoon of systems thinking related work, **1-2-3 GO!** works well as a quick opening, or awareness exercise. As with **Hands Down**, and the **Mind Grooving** exercise, this exercise can also fall in to the "gotcha" category and cause some participants to feel embarrassed or angry. This is not conducive to learning, so go lightly. The facilitator's actions mislead the group until they discover "the trap." For this reason, we only go through two or three rounds of **1-2-3 GO!** and laugh along with the group as they find themselves falling into the trap. With caution and a light touch on the part of the facilitator, this simple exercise can help participants focus more attentively on what is being said during the duration of their time together.

To Run This Exercise

Min: 3, Max: any number

5 to 10 minutes (depending on length of debrief)

No requirements other than enough space so everyone can see the facilitator

None

set-up None

Instructions

1. 2. 3. 4.

Step 1: Ask everyone to stand where they can see you, with about three feet of space between each other.

Step 2: While you demonstrate holding out your hands as if you were going to clap, ask participants to hold their arms out in front of them. Say, "Now I am going to count slowly to three and then say 'Go!' When I say 'Go!', everyone should clap their hands together."

Step 3: Slowly count "1-2-3," then clap your hands together loudly, pause for one second, then say "Go!"

When you clap your hands together almost everyone will clap their hands together, not waiting until you say "Go," as they were instructed. Pause a moment, and let everyone realize what happened.

Step 4: Repeat your instructions and then try the exercise again. Don't concentrate attention on those individuals who still clap prematurely, for they are usually quite embarrassed.

Debrief

Keeping in mind that **1-2-3 GO!** is an opening, awareness-generating exercise, the debrief tends to be light and quick. There are many different points you can make; here are just a few.

Setting the tone for the day
If you've designed an experience for the group which is very experiential and interactive by nature, you can explain that **1-2-3 GO!** demonstrates important features of the approach you will be using through out the rest of the session

Examining our automatic responses or "hard-wiring"
In **Steps to an Ecology of the Mind,** Gregory Bateson proposes that "habit is a major economy of conscious

thought...The very economy of trial and error which is achieved by habit formation is only possible because habits are comparatively 'hard programmed.' The economy consists precisely in not reexamining or rediscovering the premises of the habit every time the habit is used."

We see our automatic response in this exercise when we clap. The simple question here is: where else do we have "hard programmed" responses? Where do they serve us? Where do they not? How might we "unbundle" our thoughts and actions incorporating space for real-time reflection?

The Power of Listening

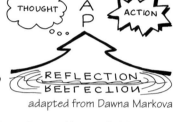

adapted from Dawna Markova

How many times have you sat down with a team to tackle a difficult organizational issue, only to spend the majority of your time working through team member's perspectives of what the problem is? A key to gaining a shared sense of current reality is effective listening. It seems simple, but good listening skills make for a better systems thinker. When you listen so you are able think and act more systemically, you:

- listen to understand rather than to debate (win-lose)

- test your own assumptions rigorously and look for disconfirming data

- reflect in real-time. Wedge in a pause for reflection between your thoughts and actions, to more often have your thoughts as opposed to being had by them

This kind of listening is vital, particularly when creating causal loop diagrams or computer simulation models. Through effective listening and inquiry, you will be better able to make the distinction between variables that are objective or empirically based.

In systems, action originates within policies. Policies

---

are procedures for taking into account a variety of information and then deciding what to do. An analyst trying to understand the system (and the policies in it) naturally talks to people in the system and asks them what motivates their behavior or how they influence others.

The exercise points out the great importance of nonverbal communication in determining what happens at the policy points. Don't attend only to what people say, look at what they do. Another point made by **1-2-3 GO!** is the role of unconscious habit, nonrational actions, in the behavior of systems.

If you are a manager in a system, it is crucially important that your actions be consistent with what you are telling people.

Here are a few questions to explore:

> What are the ways your behavior might send confusing signals to your team partners?

> Would it ever be useful to say one thing while acting as if you want something very different?

If you plan to facilitate another exercise after this opener, explain that you just illustrated for participants a new way of looking at things and learning. Point out that **1-2-3 GO!** illustrates the importance of a broader perspective when working to understand the behavior of a system. Anyone who simply reads a transcript of what the facilitator said during the exercise could not understand the point (since they would be unaware of the facilitator clapping his hands early). One must combine the verbal with the body action in this case. Stress the value of looking at things from a different perspective when learning about systems thinking, which involves a different way of looking at how things happen.

Dog Biscuits & See Saws

Most of the system tools we have, such as archetypes and casual loop diagramming, are "neck-up" tools, calling on our ability to think about the systems we are in—thus the term "systems thinking." Before we can use these tools, the challenge is to access our abilities to sense and feel that there is a systemic structure operating. Different systems feel different. Working and living in a dynamic growth structure, for example, is a different experience than living within a system that is in balance. Each has its own emotional "signature" which we can learn to recognize.

Normally we sense the presence of a system by recognizing that important variables are behaving in characteristic patterns, by observing similarities to another system we know or to an archetype, or by finding that the participants in the system collectively have a rich understanding of the causal relationships in the system.

This exercise develops people's ability to "sense" systemic structures by putting them in a playful, physical balancing system using common (but unusual) materials. Once they've learned to recognize the feeling in this simulation, they can more easily recognize a systemic structure when they walk into one at work or at home.

⇨ To raise awareness of the experiential nature of detecting systems

⇨ To experience and reflect on the feeling of working in a system with balancing elements

☆ Appreciation that systems can be "felt" and "sensed" as well as "thought"

☆ Insight into the way balancing systems operate

☆ Increased ability to detect the presence of balancing structures and delays in complex systems

"Dog Biscuits & See Saws." Just say the name of this exercise and the right side of the brain is piqued. When the dog biscuits and rulers are laid out and the exercise begins, you'll find it hard to pull people away from the task. This exercise was originally created by John Shibley and Stephanie Ryan as an innovative approach to awaken workshop participant's systems "sensing" muscles. You will find it works very well as part of a general systems thinking introduction, because it can be used to complement a didactic explanation of balancing structures, a building block in systems thinking.

To Run This Exercise

Limited only by materials and size of room

45 to 60 minutes (depending on length of debrief)

Tables and chairs. Participants work in teams of three, but more than one team can work at a table

One 12 inch ruler, manila file, and a dog biscuit (yes, dog biscuit) for each team. Medium dog biscuits, about 4 inches in length, work best. Each team will build a small see-saw with their ruler and their biscuit. You'll also need 15 small objects per participant—try mixed coins, an assortment of nuts (that you put on bolts) or a mixed collection of different paper clips. You can also use the small white ornamental stones that hotels use in landscaping—just be careful of the dust. Come prepared to provide paper and pens if participants won't have their own

set-up Have a table ready with stacks of manila folders, a pile of rulers, a pile of dog biscuits, and the objects in piles of 15. If people can gather their materials from either side of the table it will speed distribution

Instructions

1.   2.   3.   4.

Step 1: Preface this exercise with a conventional explanation of balancing loops, explaining that the structure of a system is dominated at some point by a balancing loop. Explain that we encounter balancing systems all the time. Often the first clue of a balancing system's existence is the feeling or sensation of being pulled between competing goals. (See debriefing section for more explanation of balancing structures.)

Step 2: Have the participants form teams of three and find a place to work.

Step 3: Explain that teams will be working in a balancing system that everyone recognizes—the see-saw. Tell them to place the ruler on the dog biscuit fulcrum. Hold them up as you demonstrate. There is nothing special about dog biscuits as fulcrums except they are cheap, easy to get and work perfectly.

Two people will be "workers" in the balancing system, and the third will be observing. The goal of the workers is to pile onto the ruler as many objects as they can in one minute without the ruler touching the table. Each worker is permitted to put the objects on only one end

---

of the ruler. Teams will have five 1-minute tries, and the observer will plot on a graph the total number of objects successfully balanced during each try by his team.

The role of the observer is to watch what happens to the workers and the system. They observe for a while, and then discuss what was observed.

When you're sure that everyone understands the instructions, have the groups send someone to the table for one ruler, one dog biscuit, two piles of objects, and a manila file folder.

After the first five tries, tell the workers to put the manila folder between them. This should be held in place by the observer in such a way that makes it impossible for either worker to observe the end of the ruler on the other side. The participants can also make the file folder into a simple v-shaped tent: have them tear out a piece in the center of the folder, so it can be placed over the see-saw. Each worker should be able to see his or her end of the ruler but not the opposite end.

Variation

Start the first try by having both workers close their eyes while the observer tells them what to do. On the second try, they work with their eyes open and the observer silent. The third try, have them work with the manila folder as a barrier. This allows insight into hierarchical management, delays, and learning.

Debrief

Have the teams debrief in threes first.

You can post the following questions on a flip chart or an overhead to stimulate conversation.

> What did it feel like to work in this balancing system?

> What happened? What did you each observe?

> How do the workers' observations differ from the observers'?

> How did hiding the workers' actions change the experience? Why?

Focus the large group debrief on the question, "What do you know about balancing systems?" Try to solicit a balance of thinking oriented comments with feeling and sensation oriented ones. Record their comments on the flip chart.

You might follow with the question, "If I feel I'm in a balancing structure, what do I do?" Ask them to identify where there is leverage.

If the group is new to the concept of balancing structures, this is a good time to review the basics:

Balancing structures are goal-oriented and generally stabilizing. They resist external change in one direction by producing internal 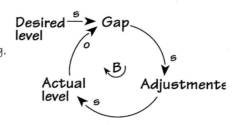 change in the opposite direction. The basic structure of a balancing loop involves a gap between the goal (or desired level) and the actual level.

As the discrepancy between the desired level and the actual level increases, corrective actions adjust the actual level until the gap decreases. In this sense, balancing processes always try to stabilize conditions.

Daniel Kim explains: "The human body consists of thousands of stabilizing processes that keep our temperature steady, our oxygen level adequate, and ensure that our body's food and water requirements are met. Organizations also have balancing processes that try to keep supply level with demand, work force comparable to work requirements, and inventory levels in balance."

But if our objectives are contrary to the implicit goal of the system, our efforts to create change will be frustrated. After sensing and recognizing a balancing process, the key to managing it is identifying the goal of the system. Most often the goal is implicit, making it difficult to realize that a balancing loop is at work. But for many, it may be more possible to sense that a balancing system is at work.

There are tell tale signs of balancing structures. Kristin Cobble, a consultant in the field of organizational learning, says, "When management goals in an organization are different than the system's goals, it feels like you're fighting the system. If you are really having to push for something to happen, giving lots of cheerleading speeches, working long hours and then, after all that effort one of the cheerleaders leaves the company or you leave on vacation and the system goes back to the way it was—you are working against that system's balancing structure."

"One example of this can be found in organizations with older product lines. Sales go in cycles due to fiscal years, weather, customer preferences, and related factors. The tendency is to push hard to get numbers up through increased sales and marketing efforts, rather than identifying the natural fluctuations of the system and exploring such possibilities as creating value added services or developing a new product line during the down turns of the cycle."

To stimulate participants' sensing abilities, you can ask a simple question: "How did this exercise feel?"

Some responses you might hear:

"We naturally did not talk. Although this was a silly activity, we quickly became engaged and focused— keeping the balance takes on a life of its own."

"It was harder, when I couldn't see what my team mate was doing."

"I felt like we were at odds with each other but trying to work together at the same time."

In this exercise, the barrier (the manila file folder) adds a perception delay to the system (the dog biscuit and ruler) showing the powerful influence of perception delays on performance. To connect this to a group's work experience, you may want make the point that often in real balancing systems you can't see the actions of the other workers, only the consequences of those actions. The group will concur that it is harder to keep the system "balanced" with the added delay.

Generally, the more delays in a system, the more oscillations, the more "extremes" you experience. After experiencing this exercise, one participant said: "This is what it feels like to be in a balancing system, feeling like you're in a competing and collaborating mode, all at the same time."

Dog Biscuits & See Saws relates directly to learning, especially if you build into the instructions an opportunity after each round for the team to discuss their performance and ways to improve it. The learning situation is quite similar to the **Moon Ball** exercise.

Source: John Shibley and Stephanie Ryan

Frames

"Seeing depends on knowledge
And knowledge, of course, on your college
But when you are erudite and wise
What matters is, to use your eyes."

Ernst Bombrich,
The Story of Art

Great folk wisdom is captured in a body of Iranian folk
tales about an itinerant preacher, Nasreddin. In one
story, Nasreddin frantically searches for something
under the light of a lamp post in the dusty street
outside his home. A kind neighbor comes by and asks,
"Mulla, what have you lost?" Nasreddin replies, "I have
lost my keys." The neighbor, being the good person he is,
gets down on his hands and knees and begins to search
with Nasreddin through the dust. After a long time,
the neighbor says to Nasreddin, "Mulla, are you certain
you lost your keys here in the street?" "Oh no!" says
Nasreddin, "I lost them in the house." "If you lost them in
the house," says the neighbor, "then why are we looking
for them under this lamp post?" "The light is better
here," Nasreddin replies.

Like Nasreddin, we often search where "the light is better." When we have problems, we look for their roots in the data that are most easily accessible to us. In the language of systems thinking, we say that people look for "proximate causes"—that is, causes that are close to the problematic symptoms in both time and space. But the causes of difficult behaviors typically lie (1) far away from the physical location where the actual behavior is observed and (2) far back in the past.

When we search for some basis to choose among alternative approaches or policies, we tend to search where the light is better. We look only at what the policies will do for us, here, and soon. But the important consequences of what we do will typically occur far away from where we take the action, and years—or decades—into the future. Debates about the appropriate policy to adopt in relation to genetically modified seeds, or CO_2 emissions, or the exploitation of oil reserves, or the protection of endangered species, all illustrate this tendency.

If we hope to understand our predicament and to find more effective solutions, it is often essential for us to consciously reframe our dilemmas, to redefine their boundaries. "Sounds simple!" you say? The challenge is that we are often unaware of how we are framing a problem. Therefore, we find it difficult to change the frame, even if the old perspective is keeping us from understanding and addressing our troubles. When we are under pressure we tend to focus even more on what we're seeing through our frame, and pay less attention to the frame itself.

This simple exercise shows us how to have our frames as opposed to being had by them. It also provides a simple but powerful metaphor: the thumb-and-finger circle that denotes a frame. Our workshop participants have found this metaphor very useful when they're conversing with others about some particularly complex issue. Simply holding up a hand with the symbol of a frame reminds everyone present to reconsider whether the prevailing choice of space and time is the most appropriate one for the problem under discussion.

⇨ To illustrate the impact of choosing different frames or points of view when defining a problem and seeking solutions

⇨ To encourage participants to try on different perspectives when diagnosing problematic system behaviors or when designing ideal systems

☆ Openness to experimentation with different perspectives when defining a problem

☆ Awareness of the impact of our chosen time horizon and geographical boundary when we conceptualize a problem

☆ Objectivity about our own paradigms

☆ A willingness to realize that the way we view something might not necessary match the way others view it, or might not be the best way to view it

It's impossible to pay attention to, gather, and analyze all the data available to us moment by moment. We have to filter the information that streams in through our many senses. We quickly learn how to use conceptual frames, or paradigms, to decide (often unconsciously) what we can ignore and what's most important. For example, when we go to the grocery store in need of milk and eggs, we most likely ignore signs advertising sales on other products such as cereal and fruit. Often we take frames from others. In these cases, such frames may better serve the purposes of the person or organization that provided them than they serve our own needs. As a shrewd leader once observed, "I don't care who gets to vote, as long as I'm the one who selects the candidates."

Most societies have several common frames through which people view reality. These frames include religious thought, economic theory, natural science, or political ideology. We can also think of these frames as paradigms—filters that direct our attention to specific forms of data, predispose us to specific theories of causality, and focus us on established kinds of problems and policies.

Any paradigm—no matter what it is—has several critical aspects:

- An implicit time horizon; that is, the length of time over which we consider information about the issue

- A geographical boundary that defines where we look for costs and benefits of alternative policy options

- Causal links that are presumed to be important. For instance, many economists disregard feedback from the environment in advocating their favorite policies. Many environmentalists, for their part, disregard the impact of the price system when arguing their own viewpoints.

Typically, we start wrestling with problems without first figuring out which frame might be most useful. A research experiment conducted some years ago by NASA offers an apt example. In 1978, NASA launched the satellite Nimbus 7 into the stratosphere to gather long-term data on significant atmospheric changes high above the Earth. However, the people who designed the experiment were working under an unexamined paradigm. They assumed that they would not have to measure ozone concentrations, because they believed that such concentrations did not change. Consequently, they programmed the computers on board the satellite to ignore information about ozone levels. Therefore, although the satellite did sense changes in ozone levels, the data were not transmitted back to Earth. If the experiment designers had operated from a different paradigm, we all would have learned much earlier about the grave damage chlorinated hydrocarbon chemicals were causing to the Earth's ozone layer. [1]

Frames become especially important during times of major change in the world around us. If we're not in the

[1] The period during which scientists were monitoring low ozone readings and yet not "seeing" them is described well in Paul Brodeur's "Annals of Chemistry: In the Face of Doubt," *The New Yorker*, 9 June 1986, p. 71.

habit of changing our frames, we may inadvertently maintain an old one long after it is no longer relevant. For example, managers often continue to monitor information that used to matter to the future of their organization but that no longer does. Meanwhile their competitors—who have moved more quickly to change their frame—begin outpacing them.

A classic illustration of this is offered by the railroads. Initially the train operators had a monopoly on the long-distance transport of freight. But then better highways and larger trucks made road transport a viable option. However, the railroads were slow to recognize the threat. For many years they continued to focus on data about "total tonnage" (one ton carried one mile). They should have monitored the fraction of all freight carried by rail. The first parameter continued to go up, long after the second started declining. By focusing on the data that used to be important, the railroad managers delayed for a decade or more any efforts to counter the competitive threat posed by trucking.

The U.S. government's efforts to fight the drug problem in the United States by sending military units to foreign drug-producing areas is another example of a flawed frame. Solutions to drug problems most likely involve social, cultural, and political changes inside the United States. As the old adage goes: "When your only tool is a hammer, everything looks like a nail." In this case, when your most effective tool for influence is your military, everything looks like an international war. Indeed, we call the government's effort the "War Against Drugs."

SUNY Albany professor George Richardson points out two kinds of frames, or boundaries—temporal and geographical—that people can choose when exploring systemic problems.

Geographic or Spatial Frame

This boundary defines the physical area over which we think people, organizations, and natural systems will be affected by the actions we take. If we adopt a narrow geographic frame, we may pay less attention to consequences that occur "over there" than we do to events that occur in our own back yard. Nations that ship their toxic waste to other, typically poorer, countries are operating from a geographic frame that ignores damaging effects of their actions outside their own borders.

Temporal or Time Horizon Frame

This boundary defines the interval of time over which we care about the costs, benefits, or results of the actions we are considering. For example, one hour, one week, one year, 10 years, 100 years.

Almost everyone gives less attention to costs and benefits that will occur in the far future than they do to those that are immediate. Economists have even coined a term—"discount rate"—to designate the extent to which we reduce our concern for future consequences in comparison with current events. If you have a high discount rate, you ignore information about consequences that will manifest themselves more than a few years into the future. This frame is particularly evident, and damaging, among elected politicians, but we all suffer from it. If people got cancer immediately from smoking, few people would become addicted to cigarettes. But in this case the consequence lies decades in the future, so most smokers ignore it and opt for the immediate pleasure of smoking.

As Richardson explains, the time-horizon boundary also has a moral dimension. For example, if you're thinking about petroleum in a one-year time frame, you may focus on price and supply. But if you take a two-hundred-year time frame, what Elise Boulding calls "the 200-year present,"[2] you cannot afford to ignore issues of pollution and of unequal quality of life among generations.

Other Frames

Other frames exist besides time horizon and geography. For example, politicians may use party affiliation as a frame. Information provided by a Democrat is going to receive much less attention than information provided by a Republican if the listener considers him—or herself—a member of the latter party.

To Run This Exercise

 Any number

 Most people use this exercise as a simple illustration within a longer and more substantive discussion of frames or boundaries. In that case, you can present it in 5 minutes. If you wish to use **Frames** as the basis for a more extensive discussion, allow 15–30 minutes.

 Participants simply sit in place, so no special space is required. The exercise does require everyone to look at a specified object that is at least 5′ away.

[2] Elise Boulding is professor emeritus of sociology at Dartmouth College, author, and peace activist. Through her imaging workshops, she helps others develop a framework that extends their view of the present. For Boulding, "The present as defined by the year, the decade, or even the quarter century is too small for an adequate grasp of significant social processes." Through her notion of "the 200-year present," Boulding has developed a compelling visioning exercise. Workshop participants imagine one hundred years into the past and one hundred years into the future. (Source: Learning to Experience Time: Interviews with Modern Systems Thinkers unpublished manuscript, Linda Booth Sweeney, 1999, Harvard Graduate School of Education.)

Each participant needs an aperture, about 1"–2" in diameter, through which they can look at different objects. Players can create this hole simply by making a ring with their thumb and index finger. The advantage of this approach is that you do not have to disrupt the flow of the discussion by passing out materials.

More important, this finger hole easily becomes a metaphor for each player's conceptual framework. After you introduce this metaphor, you will often see participants making the symbol during regular conversations, when they wish to ask about or illustrate some point related to a collectively held assumption or an individual's point of view.

Players can create a more elaborate hole by tearing out the center of a sheet of paper or of a 3" by 5" card. Have each participant take out a piece of paper. Ask them to tear a 1" hole, more or less round, in the middle. Here's an easy way to do this: Fold the paper once in half and then in half again the other way. Now tear a 1/2" piece off the corner that contains no edges of the paper. Unfold the paper, and voila! You have a sheet with an aperture in the middle.

Or you can pre-cut a 1" x 1" hole in 3" x 5" pieces of cardboard ahead of time and distribute them. The advantage of this approach is that participants have a physical object to exemplify the concept of a frame. This may make it easier for them to understand the central lesson.

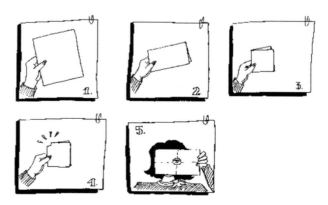

Give each participant his or her own frame, if you're using paper or cardboard frames.

Use this exercise to illustrate both spatial and temporal frames. Use only one of these approaches, depending on your teaching goals.

Instructions

Geographic Framing

Step 1: Ask all participants to hold their viewing holes out at arm's length.

Ask them to look through the holes and focus on a specific object; for example, a cluster of tennis balls on a table, a poster, you, or whatever object you choose. We often ask players to focus on a pen held by the facilitator. This doesn't require any advanced preparation or extra materials, and the facilitator can easily control the movement of the pen.

Step 2: Ask the following questions, pausing for 10–20 seconds after each, so participants have time to ponder their answer.

- "What do you see within this frame?"

- "What questions could you answer with the information available to you through your frame?"

- "What professions might be interested in the data you are gathering?"

- "What actions could you take to influence the objects or the processes that you see?"

Step 3: Now ask players to bring the hole halfway toward their eye while keeping the same object centered in the hole. Ask the same four questions above, once more pausing for several seconds after each.

Now ask players to bring the hole as close as they can

to their eye while keeping the same object centered in the hole.

Again ask the same four questions, pausing for participants to form a mental answer to each. Point out that the object that each player viewed through his or her hole remained exactly the same during the three different inspections. Explain that you asked the same questions with each inspection, then ask participants whether their responses to the questions changed. Solicit actual responses from the audience. Ask, "Why did your answers change? Which do you think was the best perspective? Which one did you prefer?" Of course, no perspective is intrinsically better than another. It depends on the goals and the questions of the person who is looking through the frame.

Temporal Framing

Step 1: Once again participants will look through their hole three times at one object. But this time, they keep the hole as close as possible to their eye during all the different inspections and only change the amount of time during which they observe the object through the hole. Ask them to choose an object in the room and look at it for a total of 60 seconds. Tell them in advance to that they'll need to record what they see after looking at the object for 1 second, for 10 seconds, and for the entire minute. Then ask the same four questions you asked above.

Again, the object stays the same. But because the time frame changes, players gather different information, generate different questions, and experience different responses with each of the three frames.

You can use this exercise to introduce the important concept of time horizon, the period of time over which a person might consider a certain behavior of interest. Here's another way to illustrate the concept of time horizon: On a flip-chart poster sheet, draw a 2′ x 2′ graph of a generic commodity price; for example, the price of pork or copper over time. The line should have short-term noise component, a 3–7 year oscillation, and a long-term (for example, 30 years) upward sweep.

Step 2: With a large piece of paper, cover most of the chart, so that participants can only see the short section of random noise. This is a time horizon of weeks or months. Ask them what they think will happen next to the price, and what is causing the changes they see. In the case of copper, short-term price changes could stem from strikes that interrupt production or weather that interferes with shipping.

Frame Graph

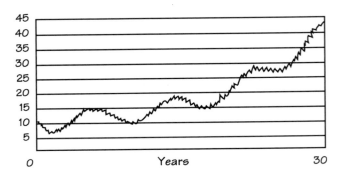

Step 3: Then reveal a time horizon of 10 years—long enough to see a complete cycle. Again ask what is going to happen next, and what is causing the changes. Long-term production cycles in copper might occur when low prices stifle investment in new mines. This reduced investment eventually renders capacity insufficient to produce all the metal that consumers desire. Thus prices rise. Then producers make investments again. After several years, production capacity increases so much that there is an oversupply. Prices fall again, and the cycle starts once more.

Step 4: Now reveal the 30 year time horizon. Ask, "What's happening? What do you think is causing these patterns?"

Long-term growth in price could be caused by depletion of high-grade ores, more stringent and costlier environmental standards, or inflation.

The point of the commodity exercise is that the same variable in the same system—the price of the commodity—looks very different depending on viewers' time horizon. No single time horizon is intrinsically better than another; choosing the best time horizon requires one to be very clear and explicit about one's goal in seeking information. A commodity trader will likely be interested in price changes over days, weeks, or months. A purchasing agent for a company that uses large quantities of the commodity will be interested in price changes that take place over months or years. Someone building a new mine needs to know what prices will look like over the next 10 years.

If time permits, participants can get a lot more from this exercise by splitting into groups of 2 or 3. Ask them to discuss where they've seen people using different time horizons for the same variable in their own organizations: for example, number of employees or students, level of inventory, motivation of the staff, price of the company's stock.

Variations

Variation 1: Combine **Paper Fold** and **Frames**. Have people look through their hole using the **Paper Fold** process for three different time periods: 1 second, 5 seconds, and 1 minute. Ask them to draw a behavior-over-time graph for the thickness of the napkin for each of the three frames.

Variation 2: Ask one-third of the group to look through their holes at the left front corner of the room, one-third to look at the right front corner, and the remainder to look straight ahead to the middle front of the room. All of them should start with their holes at arm's length and then slowly bring them toward their eyes. Ask them to describe what they see in a few sentences. In this variation, observe that initially the three groups all thought they were looking at totally different things. But by the time they had brought their frames of reference back to their eyes, they were all looking at the same thing: the entire room. This is the goal of systems thinking—to enable individuals to create frames that show the commonalities in what seem to be very different phenomena.

Debrief

In addition to the group discussion that will follow each variation of this exercise, you can encourage participants to consider how they usually frame a problem. Point out that we often fail to notice how we've framed a problem, and simply skip right to problem solving. You can also explore the various ways "framing" occurs within different fields. For example, academics typically set the frame first, and then analyze problems. As a result, an economist and a biologist will tend to look at human hunger from very different points of view. Effective managers experiment with different frames until they find one that seems to (1) offer a valid explanation of their problem and (2) lead them toward useful and enduring solutions.

Here are some suggested debrief questions:

- Is the frame more important than what you see through it? Why or why not?

- How can the members of a team determine if all of them are using the same perspective on an issue?

- How can a team ensure that all members' frames are at the same distance, or level of perspective?

- What distance or framing best captures the behavior that is of most interest to you?

- What is the link between the questions you ask and the framing you choose?

EXERCISE LINK

You may want to use **Hands Down** as a companion exercise to **Frames** to further explore the concept of framing.

Draw a large rectangle in the upper right-hand corner of a flip chart. Explain that you are going to show players a number using a code they must decipher. Then make random markings inside the rectangle while displaying the number by the fingers you place on the paper outside the rectangle. Give participants several chances to guess the secret number. Then turn the flip-chart paper to a new page on which the rectangle has been enlarged to occupy the entire top half of the flip chart. Do the exercise again, but this time rest your fingers inside the rectangle. Players should quickly guess your secret code, since the number is being displayed inside their frame of reference.

resources

For a compelling visual example of the differing spatial frames on a natural system (in this case the Earth), see *Powers of Ten*, a documentary created by Charles and Ray Eames. This 10-minute video starts by focusing on a one-meter-wide square at a picnic in a park on the shore of Lake Michigan. Every 10 seconds, the spatial frame is expanded by a factor of 10, until the camera is showing the entire known universe. We have found this film to be an entertaining and powerful way to start many of our workshops. The film is sold by Pyramid Film & Video, 2801 Colorado Avenue, Santa Monica, CA 90404. Telephone: 1-800-421-2304.

You may also find it useful to show excerpts from the book *Zoom* by Istvan Banyai. In this pictorial book, an imaginary camera pulls you back from scene after scene, and you find your perspective changing from page to page. For a description of how Banyai's *Zoom* can be used to explore the concept of frames with adults and children, see *When a Butterfly Sneezes: A Guide for Helping Kids Explore Interconnections in Our World Through Favorite Stories*, Volume I in the Systems Thinking for Kids, Big and Small Series. Waltham, MA: Pegasus Communications, 2001, p. 72–76.

Postcard Stories

"Since relationships are the essence of the living world, one would do best . . . if one spoke a language of relationships to describe it. This is what stories do. Stories . . . are the royal road to the study of relationships. What is important in a story, what is true in it, is not the plot, the things, or the people in the story, but the relationships between them."

Fritjof Capra
Uncommon Wisdom:
Conversations with Remarkable People

Often we associate story telling with parables, fables, and legends—and therefore relegate it to the province of childhood. But story listening and story telling are actually essential tools for systems thinking. As anthropologist Gregory Bateson discovered, stories can stretch our minds so that we perceive significant cause-and-effect relationships in whole new ways—an important skill in systems thinking. By hearing the stories of people who have observed a system over time, we gain an understanding of that system's characteristic behavior. By telling stories to which our listeners can relate, we help them understand the causes and consequences of certain behaviors as they occur over time.

Like a computer model, different stories about the

same system can help us see that system from many new perspectives, or frames of reference. For example, the story you narrate about a system can be from a participant's viewpoint or from an outside observer's. Stories can also cover very different time periods—a few minutes, months, or a lifetime. It's easy to tell a story; it's more difficult to make sure that it contains behaviors and relationships that are helpful to the listener.

This exercise gives participants practice in imaging and describing causal links between people and events and practice in considering the impact of different time horizons on the ways that events unfold.

The exercise can be challenging. While it gives participants an opportunity to think in new ways about common events, some may not be able to take advantage of this opportunity within the time constraints of the exercise. Some of your players will be challenged and excited; some may become frustrated. So practice with small groups first, until you have gained some mastery in leading people successfully through the sequence of steps. Provide supervision appropriate to the skill level of your group members. Do not ask your groups to develop stories about closed loops of causality, until they have become skilled at describing simple links and sequences of cause and effect.

⇨ To demonstrate the difference between a linear sequence of events and simultaneous, circular causality

⇨ To show how changing our time horizon can shift our understanding of causality

⇨ To explore the many ways causality can manifest itself

☆ Facility in telling stories that describe the form and the implications of closed chains of causality

☆ Awareness of the influence that the choice of time horizon has on our understanding of causal influence

☆ Experience in identifying delays of different lengths

As you help people better master the skills of systems thinking, you'll need to introduce the concepts of causality and the graphical tools used to represent causal links and feedback loops.

Most of these concepts and tools are relatively easy to explain. Everyone witnesses history every day, and they generally interpret it as a sequence of events, each caused in some way by events that preceded it, and each causing other events in turn. But feedback loops are not a sequence of events happening one after the other around the loop. They are a set of elements that mutually influence one another all the time. Comprehending this requires a big conceptual leap for some people. We have found that this exercise, combined with **Living Loops**, can help participants make this leap.

To Run This Exercise

 Any number of participants may join in this exercise. But keep in mind that it takes at least 3 individuals to sustain a useful discussion. If the group is not evenly divisible by 3, make up one or two groups of 4.

 Can vary enormously, from 5–10 minutes to an hour or more. The time required depends on participants' existing knowledge about causality and feedback loops and on your goals. We outline many more steps below than most users will ever have time or motivation to use. Try out different versions with participants and adapt the ones that you find to be most useful.

 Enough for small groups of 3 or 4 people to gather and hear themselves speak.

 You need a variety of interesting images in a physical format that people can handle conveniently. Postcards with images on them are perfect. Provide enough different images that each person can select one, and there will still be 20–30 remaining. That ensures that each participant has real choice in selecting the picture they will have to incorporate in a story.

You can find intriguing cards in most book or art stores. Look for cards that portray interesting situations about which participants can easily invent a short story. We find that black and white photographs work best. Avoid landscapes, photos of fine art, and portrait-style pictures of people in which there is no background. One of the postcards we frequently use is shown below.

You can also compile a workable (and more affordable) collection of images by searching through old magazines and cutting out pictures. National Geographic back issues can be an especially fruitful source. Use a copy machine that can reduce or enlarge, and photocopy the pictures in black and white onto card stock of uniform size; for example, 3″ by 5″ or 5″ by 7″. This makes the cards easy to sort, hand out, and collect. Lamination will make the cards more durable.

set-up Distribute the Cards

Spread the postcards out on a table face up. For a group of 20 people, you should have approximately 50 cards. Ask each participant to select a postcard. Tell them they should find an image that lends itself easily to being described in a 2–3 sentence story. When we use **Postcard Stories**, we write these instructions on a flip chart that we place near the table full of cards.

If you're using this exercise at the beginning of the workshop, ask participants to select one card before they sit down. We find that looking over the cards and selecting one is a good way to occupy early arrivals. As participants mill around the table looking at the postcards, they'll also find it easy to strike up conversations with one another.

To do the exercise, have participants divide themselves into groups of 3. If you have one or two extra people, add each of them as a fourth member to an existing group.

Prepare template paragraphs (e.g., "When A increases, it causes B to increase, because....") on an overhead slide or a flip chart, or write them on index cards and distribute a copy to each group. The important thing is to make sure the template is visible when you need it. See Steps 2, 5, 6, 7, 8 and 9 for sample templates.

Explain Feedback Loops and Provide Examples

This discussion can be brief or extended. It will depend on the level of your group members' competence and on your goals for the exercise.

Using a flip chart, black board, or overhead projector with slides, summarize cause-and-effect relationships and describe the graphical tools (e.g., loops, links, causal loop diagrams) that systems thinkers use to represent those relationships. Point out the nature of and the difference between reinforcing and balancing processes. For a summary of the basic concepts that need to be conveyed, refer to the materials cited below in the Resources section.

To explain the two kinds of systemic links (reinforcing and balancing) you might say something like: "In a reinforcing link, a change in one thing causes a change in another in the same direction. For example, if my performance improves at work, my workload may increase. In a balancing link, a change in one thing causes a change in another in the opposite direction. For example, if I increase the amount of food I'm eating, my hunger will decrease.

Instructions

1. 2. 3. 4.

Step 1: *Frame the exercise.* "Imagine that the scenes portrayed on your group's three (or four) postcards are linked causally. That is, an event portrayed or implied by one card has caused an event portrayed or implied on another card."

To help illustrate the exercise, hold up three postcards, designate one of them as the starting point for the story, then briefly tell a story involving the images on all three cards. Explicitly note the time horizon; that is, the amount of time that elapses in the story from the first event to the last. Name the kinds of links (reinforcing or balancing) involved in each phase of the story, and point out any important delays.

Step 2: *Establish cause and effect relationships.*
Ask players to pick two of their group's cards and to designate one as the cause and the other as the effect. Then each group member silently makes up a story about the cards in the form of a sentence describing a reinforcing link between the cause and the effect. Participants can use the following template to make up their stories:

"When A increases, it causes B to increase, because...."

Participants share their stories. If anyone has trouble linking their two cards in a plausible way, go over and offer some hints[1].

Step 3: *Notice the nature of the causal process.* Say something to the participants like: "Notice the different kinds of processes you are describing to explain how change in one part of the story produced change in the other. Some of you might be describing physical processes, such as 'the hand pumps faster, so the pump raises more water.' Or you might be describing psychological processes—like 'as the mother became more angry, the child became more fearful.' Also, note

[1] If you have ample time, you may choose to have players designate two different cards as a new cause and effect. Again they develop a story about a reinforcing link between the images on the cards. If this is difficult, players can exchange their cards for an unused one on the table that lends itself more easily to the exercise.

that the time that elapses in your stories can vary from practically nothing to many years."

Step 4: *Experiment with time horizons.* Participants experiment with time horizon. For the same cause and effect, they make up a story that describes something that takes place over a few minutes, a few days, several years, and a generation. Again, they can pick new cards from the table if necessary to find images that lend themselves to this part of the exercise.

Step 5: *Explore reinforcing processes.* Explore reinforcing processes in which both variables decrease rather than increase. Ask participants to tell new stories using this template:

"When A decreases, it causes B to decrease, because...."

Step 6: *Explore balancing processes.* Players tell stories featuring balancing processes, using this template.

"When A decreases, it causes B to increase, because...." or
"When A increases, it causes B to decrease, because...."

Step 7: *Closing the Loop.* Players work with all three (or four) of their group's cards. They place the cards in a circle and then practice creating stories that illustrate a reinforcing cause-and-effect loop. Ask them to tell a story which the events or people depicted on Card A influence those shown on Card B, which in turn influence those on Card C, which in turn, influence Card A. Ask

participants to stop and explore any confusing or seemingly impossible links that arise in their stories.

Offer the following (or some version) of the following as a template around which to build their stories:

"As there was more/less of A, there was more/less of B. As B increased/decreased, there was more/less of C. As C increased/decreased, it caused A to increase/decrease even more."

Again, players may find it useful to change the order of the cards, or to pick different elements from them to include in the story. Also, participants should create two kinds of stories: one that takes place over a very short time horizon and one that plays out over many years.

EXERCISE LINK

At this point, you might choose to conduct a quick session of the **Living Loops** exercise.

Step 8: *Further exploration.* If you have time, you may want to explore an exponential growth and collapse using the curves illustrated below.

Exponential growth. Explain that this behavior results from a collection of reinforcing links that amplify an initial increase. Cite examples, possibly of industrial growth. "As industrial capital increases, output increases because companies have more capacity to produce. As output increases, investment goes up, because investment is generally a fraction of output. As investment goes up, industrial capital increases even more." Or give the example of an argument between two people. "As person A gets angry, he express more criticisms about Person B. The more criticisms Person B hears, the angrier he gets. His anger causes him to express more criticisms of Person A, who, in turn, becomes even more angry."

Exponential collapse. Explain that this behavior also results from a reinforcing process, but one that causes rapid decline rather than growth. Cite examples to illustrate, such as soil erosion and hunger: "As food supply goes down, the use of sustainable agricultural practices goes down, because people feel compelled to use the soil even more aggressively to produce enough food. But as use of sustainable practices goes down, the fertility of the soil goes down through erosion. As the fertility of the soil goes down, food production goes down even more."

Now, as with Step 7 ask participants to lay out their cards in a circle and create a story that illustrates exponential collapse. They can use the following template paragraph:

"A was decreasing. As A decreased (got smaller), B decreased, because of.... As B decreased, C decreased, because.... As C decreased, it caused A to decrease even more."

Step 9: *Consider multiple causal loops.* Now show examples of balancing-process behavior, like the graph shown below. Explain that this kind of behavior results from a combination of balancing links, whereby a change in one thing creates change in the opposite direction in another thing. When these links connect in a circle, or loop, the resulting behavior seems to want to stabilize at a certain level. Cite examples, such as money in a person's bank account: "If the money in a person's bank account is less than what the person wants in the account, the person feels pressure to make more money. As a result, she'll work harder and (hopefully!) earn more income. As her income increases, her bank-account balance increases, too. The feeling of pressure to make more money then, declines. Both the amount of money in the account and the pressure that the person feels to make more, oscillate slightly over time but hover around the same level." If the graph below were used to illustrate this simple bank account example, the goal would represent the level of bank balance desired. The rising curve would illustrate the amount of money actually accumulated in the person's personal bank account.

GOAL

Ask the group to put their three (or four) cards in a circle and make up a story that illustrates a balancing process in which the circle of causation acts to bring some feature, illustrated by one of the postcards, up to its desired level.

The precise sequence of interactions can vary so long as the overall affect of the loop is to counteract the initial change. One possible template for this would be:

"A was increasing. As A increased (got bigger), B increased, because of.... As B increased, C decreased, because.... As C decreased, it caused A to decline."

Variations

Variation 1: Exploring Time Horizons
If you want to emphasize the importance of time horizons, consider the following variation:

On a flip chart or overhead projector, show three different time horizons: 1 hour, 1 year, and 50 years.

Each group member then makes up a story linking the group's three postcards. One person tells a story that takes place within an hour; the second person tells one that happens in a year; the third, a story that spans 50 years. Each story should be 5–10 sentences in length.

The point here is that the same data (a postcard picture) can be interpreted very differently when the time frame changes.

Variation 2: Same Structure, Different System
Have participants switch cards with members of other groups and then create a new story line using the same reinforcing or counterbalancing story line that they used previously. Point out that the new story may have the same structure as the old one (reinforcing or balancing links). However, the system described in the story is different because the components of that system (the images in the cards) have changed. Ask participants where in everyday life they have seen very different

systems exhibit reinforcing growth, reinforcing decline, or counterbalancing behavior.

Debrief

Ask a few people to volunteer to come to the front of the room with their group's three cards. They hold up the cards and briefly tell their story. If you use Variation 1, you might ask them to share the story from the perspective of three different time horizons.

Voices from the Field

Carol Ann Zulauf, professor of organizational studies at Suffolk University, has used **Postcard Stories** with her graduate school classes. "The students love this one," she explains. "I remember their sitting around, each with a different card, and making up connections as they went along. I also remember one student commenting that a slightly different perspective could be had depending on where you started with the cards. This exercise was easy to administer and easy for participants to follow."

resources For a simple illustration of a narrative that describes feedback, see the well-known children's book *If You Give a Mouse a Cookie*, by Laura Numeroff. In this story, a simple reinforcing pattern emerges: If you give a mouse a cookie, he will want a glass of milk. After a series of additional events, he will eventually want yet another cookie. This story helps kids practice tracing cause-and-effect relationships to see how the event (giving a mouse a cookie) actually feeds back on itself.

For more on how stories can help children explore systems thinking concepts, see *When a Butterfly Sneezes: A Guide for Helping Kids Explore Interconnections in Our World Through Favorite Stories*, Volume I in the Systems Thinking for Kids, Big and Small Series. Waltham, MA: Pegasus Communications, 2001.

In addition, Pegasus Communications offers two resources designed to facilitate story telling about complex systems. These include:

- *Selecting Variable Names for Causal Loop Diagrams*, by Kellie T. Wardman, The Systems Thinker Vol. 5, No. 6, 1994 Pegasus Communications, Inc., Cambridge, MA.

- *Designing a Systems Thinking Intervention: A Strategy for Leveraging Change*, by Michael Goodman et al., 1997 Pegasus Communications, Inc., Cambridge, MA.

Paper Fold

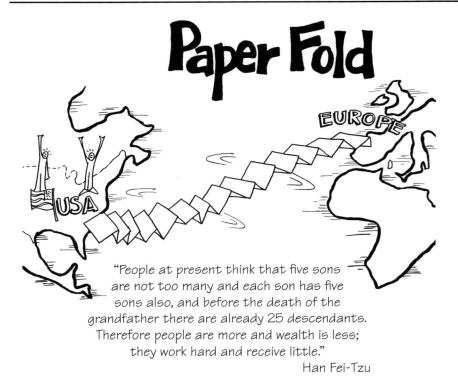

"People at present think that five sons are not too many and each son has five sons also, and before the death of the grandfather there are already 25 descendants. Therefore people are more and wealth is less; they work hard and receive little."

Han Fei-Tzu

Ancient Chinese philosopher Han Fei-Tzu described a quality of exponential growth that can be tricky to understand: The larger the quantity (for example, number of seeds, amount of money in a bank account, number of children in a family), the greater the rate of growth of that quantity. Thus, as a consequence of this reinforcing process, the greater the quantity in the future. These reinforcing loops lie at the heart of common examples of exponential growth, e.g., compounding interest, rising productivity, arms races, and population growth.

As Han Fei-Tzu revealed when he spoke about the hidden consequences of population growth, we are limited in our perceptions related to exponential growth.[1] So why do most of us profoundly misunderstand exponential growth? One answer may be that most expansion and change in our environment is essentially linear in nature, therefore most of our experience is with linear growth, not exponential growth. Many of us rely on our experience with linear systems when we estimate the numbers or the behavior that will result from exponential expansion. For example, a linear process of expansion might add 10 acres to your land holdings each year. An exponential process would add 5 percent of your current holdings annually. A linear process occurs when driving the car: you add, for example, 50 miles per hour to the distance traveled. An exponential process would, for example, double the speed every hour. Many of the goals we express imply exponential growth. For example, "I want my salary to increase by 5–10 percent per year." Or, "I want word-of-mouth to help sell my new book." Or, "I expect my stock portfolio to gain in value at least 20 percent per year." Or, "Each year I want a computer that is 30 percent faster than the year before."

The **Paper Fold** exercise is a fun, simple and extremely portable way of engaging people to think about the power of reinforcing processes and exponential growth.

⇨ To help participants experience exponential growth in a tangible way

⇨ To increase understanding of shifting dominance

⇨ To practice drawing behavior over time graphs and interpreting causal loop diagrams

[1] Extensive research has shown that regardless of mathematical training, most people significantly underestimate the power of reinforcing processes and exponential growth. For a discussion of this research, see John Sterman (1994) *Learning in and about complex systems.* System Dynamics Review 10(2–3), p. 291–330.

☆ Awareness of how positive feedback loops manifestthemselves in everyday life

☆ New insight into the complex and/or surprising behavior that can result from the simple structures that produce exponential growth

Paper Fold graphically illustrates the power of reinforcing processes. You may also use it to explore a related phenomena, that of *shifting dominance*. Because such a system will produce extreme rates of change very quickly, any system that is dominated by reinforcing processes will eventually encounter one or more limits. These limits cause balancing processes to kick in. As they do, the reinforcing behavior will level off or even reverse. This change from reinforcing to balancing is known as shifting dominance. By better understanding shifting dominance, we can more easily detect reinforcing processes in their early stages and intervene appropriately before they spiral out of control in our lives.

When we are struggling to help our clients or audience understand the behavior of some reinforcing loop that lies at the heart of a relevant issue, we often find it useful to take five minutes to do the **Paper Fold** activity. We like this exercise in part because of its simplicity and portability. If presented in the spirit of inquiry, exploration, and playfulness, **Paper Fold** can help participants confront their own misperceptions about causality and exponential growth in a nonthreatening way. When **Paper Fold** is followed by several thought experiments (also described here), it raises participants' awareness of the dynamics of exponential growth that often lie hidden from view in our everyday experiences.

Cautions

Like every exercise in which participants learn from their own mistakes, **Paper Fold** should be used with some care. Make it clear that mistakes in estimating behavior are not a reflection on any participant's intelligence. Rather, they stem from the difficulty we all have in projecting the behavior of reinforcing loops.

To Run This Exercise

 Any

 5–15 minutes, depending on whether you will introduce causal loop concepts

 Any

 1 small cocktail napkin or paper-towel square per participant. One bed sheet for use by the facilitator (A bed sheet works well when you have a large group but it is optional. You may choose to use a piece of paper instead)

 If you'll be conducting this exercise at the beginning of a workshop session, put out the napkins or paper towels ahead of time. Typically, we put one sheet of paper under or on each participant's chair. If you're conducting the exercise later in the session, hand out the paper when it is needed.

Instructions

Step 1: *Invite the group to carry out the following actions while you demonstrate with your own piece of paper.*

Take the napkin (or paper-towel square). Fold it in half, fold it in half again, and fold it in half again. Now fold it in half a fourth time. After four folds, the wad of paper is about one centimeter or 0.4″ thick. Hold it up edgewise, so that participants can see the thickness. Hold it loosely in order to avoid pinching it down to a thinner cross section. Say: "Of course, you could not physically fold this napkin in half 29 more times. But imagine that you could. How thick would it be then? After four folds it is one centimeter. How thick will it be after 29 more folds?"

Step 2: *Invite answers from the group. Because the correct answer to this question is highly*

counterintuitive, most people will not know it. Many participants may avoid replying so as not to make a mistake. Or someone might say, "To the moon." If you get this latter reply, state "No, not to the moon. Not even close!"

Step 3: *Stimulate discussion about possible answers.* We suggest to our audiences a number of different thicknesses and ask them to raise their hand for the answer that seems most reasonable to them. For example, we say, "Who thinks it would be less than a foot thick? (Pause for a show of hands.) Who thinks it would be less than from the floor to my waist? How about from the floor to the ceiling? How about from here to the top of the building?" For a laugh you can ask, "Who is waiting to see which answer gets the most votes?" Then we give the correct answer: "Folded 29 more times, this napkin would be more than 5,000 kilometers, or more than 3,400 miles thick. That is approximately the distance from Boston to Frankfurt, Germany."

Variation

This exercise may also be conducted without giving each participant a piece of paper. Simply hold up a large bed sheet or table cloth in front of the room. Fold it in half four times. Show the edge each time and announce that after four folds it has grown to be about 1 cm thick. Then ask the participants how thick it would be after 29 more folds. Proceed with the explanation and the debrief.

Debrief

Most participants consider the correct answer preposterous and assume there's a trick behind it. Therefore, in debriefing the exercise, we suggest first demonstrating the math behind the answer. Use slides or a white board to show the dramatic outcome of doubling anything 33 times: 1, 2, 4, 8, 16, etc. Doubling something 29 times increases it by a factor of about 540 million. After four folds, the napkin is about 0.4" thick. Doubling it 29 more times would produce a thickness of 216 million inches. A mile is about 63,400 inches, so the folded napkin would be a little over 3,400 miles thick.

You can quit at this point, having demonstrated that the process of doubling quickly produces unexpectedly enormous numbers.

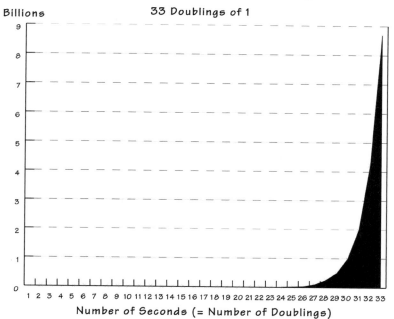

Potential Growth from Doublings

Number of Seconds (= Number of Doublings)

If you have the time and the motivation, however, you can help participants derive numerous other lessons from this simple exercise.

For example, we often ask people to draw a behavior over time graph for the increasing thickness of the napkin, assuming that they could accomplish one fold every second for 33 seconds. Then we ask where else they have seen this sort of behavior. Population growth and the spread of a cancer are two phenomena that often come to mind.

Depending on the time available, you may want to elaborate on the issue of population growth, since it's a dramatic one that piques many people's interest. You can say, "We chose to illustrate 33 doublings in this activity for a reason. Today's global population is almost 33 doublings from the first person on Earth. More than 6 billion people currently live on the planet. In other words, the thickness of a single sheet of this napkin is to the Boston-Frankfurt distance as a single person is to the current population of the entire globe."

A traditional French riddle also illustrates the surprising nature of exponential growth:

> Suppose a water lily is growing on a pond in your backyard. The lily plant doubles in size each day. If the lily were allowed to grow unchecked, it would completely cover the pond in 30 days, choking out all other forms of life in the water. For a long time, the plant seems small, so you decide not to worry about cutting it back until it covers half the pond. How much time will you have to avert disaster, once the lily crosses your threshold for action?

The answer is, "One day." The water lily will cover half the pond on the 29th day; on the 30th day, it doubles again and covers the entire pond. If you wait to act until the pond is half covered, you have only 24 hours before it chokes out the life in your pond.

The behavior in all of these instances seems counter-intuitive. We generally expect things to follow linear patterns of growth. For example the height of a pile of paper grows linearly, when new sheets are added to the top of the pile at a constant rate. With linear growth, the initial change is the same as the change later in the process. But positive feedback causes a process that starts slowly. In folding the napkin, no significant change is noticeable for many doublings. Then, although the underlying growth process hasn't changed at all, an explosion seems to occur. The 34th doubling would actually add another 3,400 miles to the napkin's thickness.

Point out that the napkin obviously does not get so large. That's because participants run into a limit. There is not enough cloth to sustain the folding process. Ask participants to continue folding as long as they can, one fold per second, and then to draw on a behavior over time graph the behavior they *actually* experienced.

The pattern they draw depicts a shift from *exponential growth* to *equilibrium*.

To help participants understand this behavior, it is useful to show a causal loop diagram of the underlying loop structure. Display the following diagram.

Positive Feedback Gives Exponential Growth

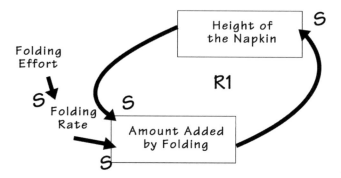

Here, R1 is the dominant loop. For a constant folding rate, the greater the thickness of the napkin, the greater the amount added by folding. As the amount added by folding goes up, the thickness of the napkin increases as well.

Actual Growth from Doubling

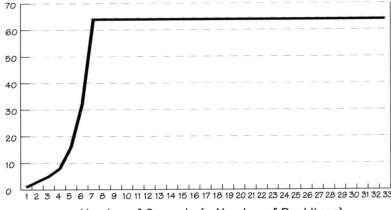

Number of Seconds (= Number of Doublings)

The exponential growth plateaus, once you can no longer physically fold the napkin. What causes this plateauing? The answer is: shifting dominance.

Shifting Dominace Stops Growth

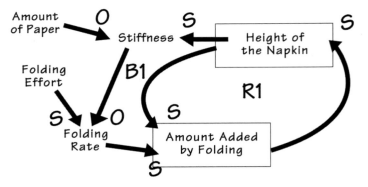

Initially, change in the napkin's thickness is influenced only by the reinforcing loop (R1). While that loop is dominant, growth in the napkin's thickness does not produce any perceptible increase in its stiffness. But as the thickness increases, the stiffness starts to increase. The resistance to folding grows until no amount of human effort can produce another fold. The balancing loop (B1) has become dominant.

Shifting dominance is an important phenomenon for

leaders from all fields to comprehend whether the fields be business, education, non-profit, healthcare, or government. When it occurs, successful policies that people have learned and refined over time no longer work; they may even become counterproductive. Management lore is full of stories about leaders who attained success by identifying and pushing on the dominant reinforcing loop that has been driving progress in their organization. But then some limit emerges—perhaps market saturation of a product or a new strategy by competitors. Because the company's data systems probably focus only on the variables in the loop that used to be dominant, leaders' control systems don't even register the change. Performance eventually falters, but management's response is to push even harder on the policy levers that used to work—to no avail. By the time indisputable evidence emerges that new loops are now dominant, it may be too late to prevent permanent damage. Understanding the dynamics of shifting dominance can help managers react to changing conditions before it's too late.

VOICES FROM THE FIELD

John Sterman, the Director of MIT's System Dynamics Group and the J. Spencer Standish Professor of Management, has used variations of the **Paper Fold** exercise for years in his system dynamics classes at MIT, with excellent results. Here are his comments:

"I usually do the paper folding exercise at the start of the term, as an illustration of the counterintuitive character of complex systems, and to motivate the need for formal simulation models to help us understand the dynamics of systems (our mental simulations of even this very simple system are systematically wrong—by huge amounts. Thus to reliably understand the implications of our assumptions we need assistance in the form of computer simulation).

"When time is shorter, I pose the problem verbally, illustrating with an actual sheet of paper, then asking for a show of hands: 'how many think the paper will be one meter thick (or less)?' Write the number on the board. Next ask 'How many think it will be between 1 and 10 meters thick?' I always give a suitable illustration of the distance involved, e.g., 'Ten meters is the length of this room.' Write the number down. Continue by powers of ten, with illustrations, until you've got most of the hands. Usually a few people have heard the problem before, or suspect you of tricking them and won't answer, so you can then ask them how thick they think the paper will be. Quite reliably the average answer will be very much smaller than the correct one.

"The **Paper Fold** exercise positions you well to discuss exponential growth in a wide variety of real situations, to identify the balancing feedbacks that might halt the growth, and to consider whether time delays in these loops might lead to overshoot and decline."

Tracy Benson from the Waters Foundation offers these observations of the use of **Paper Fold**:

"I use **Paper Fold** in my work with middle-school algebra teachers, as a follow-up to an experiential lesson. In the lesson, we hook up a motion detector to a graphing calculator and ask students to physically move so as to create different types of graphs on the graphing calculators. For instance, they create graphs of linear growth and decay, exponential growth and decay, etc. The students have to physically enact the constant rates of movement for linear growth, and the changing rates for the nonlinear, compounding growth. We then do **Paper Fold** and have the students build a STELLA model that represents the dynamics in the activity. They generate both table (spreadsheet) and reference behavior graphs.

"During any kind of lesson about exponential activity, I've also found it helpful to ask what else students see growing or decaying exponentially in our world. Answers include rumors, disease, information on a phone tree, populations (human, animal, micro-organic), fads (Pokémon, tattoos, piercing, fashions), popularity (music groups, political candidates, movie stars, etc.). High school teachers have told me that these kinds of connections help students make their own fragmented education (science, math, English, history, etc.) more interdisciplinary.

"...When we did the activity with seventh-grade students and asked how thick the napkin would be if they could fold it 29 times, some students in the class who knew the equation for exponential growth (doubling) immediately took out graphing calculators and plugged in '2 to the 29th power.' They were not at all fazed by the size of the answer, and proceeded to articulate an in-depth understanding of exponential growth. Many decided that they were going to challenge their parents at the dinner table that night and see if they could generate accurate insights using the napkin. We then launched into a conversation about the difference between doubling and multiplying by 2. That conversation led to further questions about the possibility of base numbers with exponents with exponents. It was mind-boggling."

resources Description of this exercise and related research can be found in John Sterman's, *Business Dynamics: Systems Thinking and Modeling for a Complex World* (Irwin/McGraw-Hill, 2000).

A favorite book for overall behavioral decision making is *The Psychology of Judgment and Decision Making* by Scott Plous (McGraw Hill, 1993) This little book covers all the key findings in the field of decision making. It includes a comprehensive discussion and references to the primary literature on paper folding and other related experiments on exponential growth.

Monologue/Dialogue

> "No one would talk in society, if he only knew how often he misunder- stands others."
> Goethe

If two or more people are committed to understanding the nature of a system of interest, together, they will have to surface, describe and begin to discuss the interconnections among significant, underlying assumptions. Typically, they find themselves communicating with each other about a phenomenon they have not fully experienced and do not fully understand. In these situations, group members may forget to ask questions and instead, will simply advocate their respective points of views. They may also lack the shared vocabulary and skill needed to surface and clarify important assumptions.

When faced with such messy situations, it helps to gather around a table or flip chart and begin to map out some of the group's assumptions on paper. A relevant behavior-over-time graph or a simple causal loop diagram can be worth a thousand words. But to create drawings that accurately represent the collective opinions and mental models of a group is not easy. It sounds simple, but each group member needs to know how to talk effectively. Good communication lets people surface and test their mental models and learn together. Oftentimes, different individuals focus on different aspects of the issue. When that happens, group members who engage in dialogue— two-way communication that includes both speaking and listening[1]—rather than monologue—one-way communication in which one person speaks and the other listens—have greater success. The **Monologue/ Dialogue** exercise raises awareness of the ease with which misunderstandings can crop up. It illustrates the differences between the two modes of conversation in both procedure and results.

⇨ To illustrate the differences between monologue (or one-way communication) and dialogue (or two-way communication)

⇨ To improve each team member's ability to carry out reflective inquiry

⇨ To help participants see the connection between good two-way communication and improved systems model creation

☆ Appreciation for the characteristics and the benefits of dialogue

☆ An example that participants may cite if they stray from dialogue toward monologue during problem-focused discussions

As defined by physicist David Bohm, dialogue is a profound mode of conversation that helps participants uncover and alter their deeply held assumptions. It

---

[1] As a practice, dialogue has a long history and typically involves a circle of people exploring a given topic. Each attempts to suspend his/her judgments so as to be able to inquire into a variety of perspectives. Group members work to recognize their own assumptions and to listen for the intention behind the words that are being expressed. For more on the practice of dialogue, see "resources" at the end of this exercise.

is a process through which all the participants gain understanding. The simple, quick exercise described here does not give the opportunity to explore all aspects of such dialogue. But this style of reflective conversation does require that participants accomplish the skillful interweaving of listening, reflecting, and talking. Some characteristics and effects of that interaction are illustrated in this game.

When individuals come together to "solve a problem, they bring a plethora of perceptions about the challenges and priorities at hand. Meetings can quickly degenerate into individual efforts to ensure that the others "understand" the situation (translation: that the others share the same views as the speaker!). To avoid that, we often use this exercise just before a planning session. It reveals the difference between effective and ineffective communication styles. It also gives participants an opportunity to develop rules for the conversation they are about to have. Merely carrying on a two-way conversation does not guarantee that you are in dialogue. But if you are engaged in one-way conversation, you certainly will not achieve the benefits of dialogue.

Like all didactic exercises, this one is more effective when it is delivered to participants who know they will soon have to apply the lessons they learn. So we typically introduce **Monologue/Dialogue** by referencing some planning or discussion task that the participants will face later in the workshop or when they return home. We may even give the audience the debriefing questions listed below and ask them to divide themselves up into the task groups they will be forming in the future, to deal with some important real issue. They can then discuss the questions in small groups and decide on several communication principles they will adopt when working later in their task groups.

To Run This Exercise

This exercise is engaging enough that you can use two teams up front and ask everyone else to observe. You'll need four volunteers to perform the activity in front of the audience; any number may watch.

 20 minutes

 A podium or other clear space in front of the room, so that the audience can see what the teams are doing. The audience may sit in chairs or in a circle on the floor. The space should permit you to sequester two people (out of the main group's hearing range) for a few minutes at the beginning of the exercise.

 Flip chart, or overhead projector and blank projector sheets, suitable drawing implement for either the flip-chart paper or projector sheets

 During the first phase of the exercise, a participant will need to face the audience without being able to see the activity going on behind him or her. That activity will consist of a team member creating a drawing.

Set up the easel or projection screen in a way that permits an individual to stand in front of it, with his or her back to it, and that lets the audience see the flip-chart paper or screen at the same time.

Also prepare in advance a flip-chart page or overhead projection sheet with the following illustration. Copy the same illustration onto a 3″ x 5″ card.

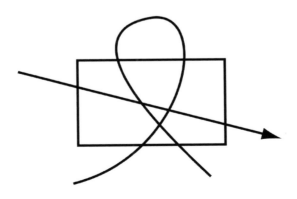

Cautions

Standing in front of a large group can be challenging for many people. When you ask for volunteers, explain that they will have to carry out a simple activity in front of the audience.

This exercise can generate some hilarity. Be sensitive, though, to team members' feelings. They'll be struggling under difficult circumstances to produce a high-quality drawing. Always be ready to offer some positive statements during the exercise, regardless of how poorly the teams may be doing. For example, "If we were offering a prize for creativity today, this would definitely win the gold!" Or "This could pass for an early drawing by Picasso—definitely a winner!"

Instructions

Step 1: *Ask for volunteers.* Ask four people to come forward and form two teams (two people in each team). Tell them that the exercise will involve drawing, but that it doesn't require any artistic expertise. Then ask each team to designate one member as the "Artist" and the other as the "Speaker."

Step 2: *Identify one pair as Team A; the other as Team B.* Ask the two members of Team B to leave the room and close the door, so they can't hear the conversation which will take place at the front of the room. Indicate that they will be invited back in about five minutes. If you have an assistant available, ask him or her to escort Team B out of the room and remain with them.

Step 3: *Set up Team A.* When Team B is safely out of hearing range, position Artist A at his or her drawing surface (the flip-chart page or overhead projector). Make sure the surface is positioned so that everyone in the audience can easily see what is being drawn. Give the Artist an appropriate drawing pen.

Position Speaker A so that he or she is facing the audience and cannot see the drawing. Give Speaker A the 3" x 5" card with the illustration on it. Make sure that Artist A cannot see the card.

Step 4: *Read the exercise instructions:*

- "Your team's goal is for your Artist to produce, as accurately as possible, a duplicate of the figure on your Speaker's card.

- The Speaker will have three minutes to describe the illustration, and the Artist will use that description to draw the figure.

- The Artist may not see the diagram that the Speaker is holding.

- The Speaker can use any words he/she wants, but isn't allowed to move his/her hands or in any other way reveal the diagram to the Artist. The Speaker can use only words.

- The Artist may not ask questions, and the Speaker must not see what the Artist is doing. The Speaker will keep his/her back turned to the Artist and the drawing during the entire exercise. Ready? Go!"

Step 5: *Make sure the rules are followed.* Take mental or written notes that will remind you of specific events or comments that you can use later during the debrief. Stop the team after three minutes. Thank them. Offer a sincere compliment about the quality of their result. Then show Artist A and the audience the illustration on the Speaker's card.

Step 6: *Invite Team B to enter the room.* Now hide Team A's work and the 3" x 5" card. Ask Team A to take

their seats in the audience, and invite Team B to enter the room. Position Artist B at the drawing surface with a pen. Position Speaker B so that he/she is facing the Artist and the drawing. Typically this means that Speaker B's back will be to the audience.

Step 7: *Read instructions for Team B. Read the instructions below. (NOTE: They're slightly different from the instructions for Team A.)*

- "Your team's goal is for your Artist to produce as accurately as possible a duplicate of the figure on your Speaker's card.

- The Speaker will have three minutes to describe the illustration, and the Artist will use that description to draw the figure.

- The Artist may not see the diagram that the Speaker is holding.

- The Speaker can use any words he/she wants, but isn't allowed to move his/her hands or in any other way reveal the diagram to the Artist. The Speaker can use only words.

- The Artist may ask questions, and the Speaker can see what the Artist is doing. Ready? Go!"

Step 8: *Again, make sure the rules are followed. Take mental or written notes that will remind you of specific events or comments that you can use later during the debrief. Stop the team after three minutes. Thank them. Offer a sincere compliment about the quality of their result. Then show Artist B the illustration on the Speaker's card. Invite Team B to take their seats in the audience.*

Place all three drawings (the original, Team A's result, and Team B's result) in plain view of the audience.

Debrief

Team A demonstrated a one-way form of conversation called monologue, whereby one person simply talks at another person. Team B demonstrated dialogue, whereby both members of the team can speak and answer questions. This exercise illustrates the importance of dialogue in two ways. First, the quality of the Speaker's instructions will be markedly higher in the dialogue situation. In the monologue situation, Artist A will often make some wildly inaccurate line on his or her drawing. It is seldom corrected. When Artist B makes a similar mistake in the dialogue situation, Speaker B will likely intervene and request some changes.

Second, the resulting drawing will in most cases be better for Team B because of the impact of dialogue.

As the facilitator, you must be prepared for the possibility that the dialogue team might actually produce a poorer picture than the monologue team. This happens about once every 10 times the exercise is delivered. If this does happen, point out that no simple rules will ever guarantee success. But ask the audience to comment on the difference in the quality of the communication in both teams. Generally the dialogue team has a much richer communication between Speaker and Artist.

You might reduce the possibility of a superior outcome from monologue by creating a more complex diagram than the one shown in this volume and by giving the teams a bit more time.

When members of a team are struggling to identify the key variables and causal links involved in a system problem, dialogue is particularly important. If your participants have had experience in drawing causal diagrams, you can ask them to reflect on that process when addressing the following questions.

Here are some tried-and-true questions to ask:

- Which method of communication did you think was more effective? Why?

- On the basis of this exercise, what rules or principles would you try to use to ensure that communication is maximally effective when you are working with others to address a task or problem?

- In your own organization, where do you see monologue taking place instead of dialogue?

- What types of organizational structures encourage monologue? Dialogue?

Here are some responses we have heard in the past:

- Effective communication is participative.

- All team members are fully engaged during dialogue.

- Messages and the means of delivering them must be tailored to the needs of team members.

- A feedback system is necessary, so that the person receiving information during the dialogue can test it and incorporate it into further discussion and decisions.

Source: **Monologue/Dialogue** is adapted from a warm-up exercise originally demonstrated to us by Dr. Elizabeth Christopher, author of several books on the use of games to teach leadership skills.

resources

Glenna Gerard and Linda Ellinor, *Dialogue: Rediscover the Transforming Power of Conversation*, John Wiley & Sons, Inc. 1998.

William Isaacs, *Dialogue and the Art of Thinking Together*, Doubleday/Currency 1998.

Peter Senge, Richard Ross, Bryan Smith, Charlotte Roberts, Art Kleiner, *The Fifth Discipline Fieldbook*, "Team Learning" p. 351–406, Currency/Doubleday, New York 1994.

MENTAL MODELS TEAM LEARNING SYSTEMS THINKING SHARED VISION PERSONAL MASTERY

Living Loops

"...[T]he term 'feedback' has come to serve as a euphemism for criticizing others, as in 'the boss gave me feedback on my presentation.' This use of feedback is not what we mean in system dynamics. Further, 'positive feedback' does not mean 'praise' and 'negative feedback' does not mean 'criticism.' Positive feedback denotes a self-reinforcing process, and negative feedback denotes a self-correcting [balancing] one. Either type can be good or bad, depending on which way it is operating and of course on your values."

John Sterman
Business Dynamics:
Systems Thinking and
Modeling for a Complex World

The notion of a feedback loop is not always an intuitive one. For people encountering causal loop diagrams for the first time, the illustrations may convey no more information than a plate of spaghetti would. Even skilled analysts often have trouble interpreting the various dynamic implications of these diagrams. Yet, causal loop diagrams can be an efficient way to represent our understanding of the feedback structures driving a system's behavior. This exercise helps participants understand, through first-hand experience, the behavior inherent in simple feedback loops.

Living Loops helps participants understand the differences between open and closed chains of causation—when there is no feedback and when there is. It also illustrates, in a physical way, the kinds of behaviors that reinforcing (or positive) and balancing (or negative) loops create. It provides practice in drawing behavior-over-time graphs and combines analysis with body movement and visual perception. Moreover, it encourages participants to hypothesize about the impact of a simple change in the sign of a link or loop, and then to test their hypotheses. This activity is a simple, effective, and engaging way to physically, mentally, and emotionally experience the two building blocks of system behavior. All learners—but particularly those who learn best through physical experience—can benefit from **Living Loops**.

⇨ To show the profound difference between open and closed loops

⇨ To illustrate through participants' own physical movements the behavior of balancing and reinforcing feedback loops

⇨ To link physical experience with intellectual analysis of behavior in closed chains of cause and effect

⇨ To demonstrate the principle that balancing loops have an odd number of negative (or "-") links and reinforcing loops have an even number of negative links

⇨ To provide practice in drawing behavior-overtime graphs

OUTCOMES

☆ A more intuitive understanding of the basic dynamics in simple feedback systems

☆ A fuller appreciation for the profound impact that adding a loop or changing the polarity of one or more links in the loop can have (that is, converting a "-" link to a "+" link, or vice versa)

☆ The realization that we can use our senses as well as our minds to perceive systems' behaviors.

CONTEXT

Organizational learning practitioner and innovator John Shibley has worked with several colleagues to create an experiential way to explore the dynamics of complex systems. We have adapted his exercise, adding many features, and we present it here as **Living Loops**. This game is an ideal complement to a lecture and blackboard presentation on the basic principles of causal diagramming.

In addition to providing a way to depict circular feedback, the exercise helps group members become better acquainted with each other. People enjoy playing together and cooperating. In many cultures, slight physical contact—such as holding hands while standing in a circle—may enhance this process.

To Run This Exercise

Number of People

6–12 as players; any number as observers

Time

About 20 minutes

Space

A room free of obstacles, in which a group of 6–12 people can stand shoulder to shoulder in a single line, or in a circle holding hands. There should be enough room for observers to watch the action and hear you, the facilitator. With groups of 50 or more, a raised platform or stage is very helpful

Equipment

1 ball or other colorful object that can be held easily in a player's hand; a loudspeaker if observers number several hundred; one 3"x 5" card per player (not per observer)

that has a large "+" sign on one side and a large "-" sign on the other; a 1 yard-long piece of string per player; a stapler or roll of clear tape; a chalkboard, white board, or flip chart

Cautions

This exercise involves gentle movement. Nonetheless, it is always wise to attend to participants' physical and psychological comfort. To participate with ease, every player should be able to move his or her arms and hands easily from below the waist to high over the head. Before you ask for volunteers, point out that the activity will involve extreme reaching and some bending. Suggest that anyone who would experience physical discomfort with these movements will make a great observer.

With small audiences (10–15) in which most people will be players, suggest that anyone who tends to experience discomfort with reaching and bending can serve as a quality-control partner. In this role, she or he will stand outside the circle, watch players' movements, and ensure that players move in accordance with the rules of the game.

If you have a large audience and many physically challenged participants, assign one to each circle of 6–12 players. You can also adapt the exercise in advance for physically challenged players. For example, if a participant uses a wheelchair, you could easily conduct the exercise with everyone sitting down.

During the exercise, it is inevitable that some players will make a mistake and move in the opposite way from what the rules indicate. They may feel embarrassed. When a mistake happens, immediately and warmly say, "Thank you for illustrating a common mistake that we all make when we're trying to interpret the signs on a causal loop diagram." Restate the definition of a positive and a negative link, and start again.

If you are running this activity for participants from a culture in which strangers, or men and women, do not comfortably hold hands, give each person a short length of rope, about 12″ (30 cm) in length. They can each hold onto one end. [1]

 To each 3″x 5″ card, tape or staple one end of a piece of string to the upper right corner of the card; tape or staple the other end of the string to the upper left corner of the card. Players can then slip the loop of string over their heads and display the card on the front of their torsos.

Move chairs and tables to the sides of the room if necessary, to give people enough space to play and observe.

Instructions

 1.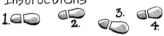

NOTES:

- The instructions for **Living Loops** are detail-intensive. Reread them several times before conducting the activity.

- Depending on your objectives, you may want to give the group an overview of the two types of feedback loops before conducting the exercise. Explain that we are often introduced to reinforcing and balancing loops through causal loop diagrams— both reading and drawing them. Point out that this exercise provides an alternative approach—that is, it lets people become these loops instead of merely drawing them.

Step 1: *Establish an open link.* Invite 5–12 volunteers to come up and serve as players.

Ask the players to stand in a line, shoulder to shoulder, facing the observers. Ask all the observers to take out a

---

[1] Tracy Benson has used **Living Loops** several times. Her participants do not hold hands or a piece of rope; they use colored hoses: "Workshop participants seemed to think that the colored hoses (green for same or positive connections and red for opposite or negative connections) may be helpful, especially for children who are visual learners."

piece of paper and a pencil or pen.

Give one card to each player. Have players put the string loops over their heads and position the cards so that the "+" signs face out. Put a "+" sign card on yourself.

Give the ball or other visible object to the person standing at the far right end of the line, when viewed from the audience. He or she holds the object in his or her left hand. Walk to the other end of the line and get in line yourself, facing the audience on their left.

Explain that every player's left hand is going to be "active." Ask players to clench their left hand into a fist and hold it out at waist height. Do that yourself.

Explain that players' right hands are "passive." Ask them to rest their right hand lightly on the fist of the person to their right. Let your own right hand hang freely.

NOTES: If you decided ahead of time that people would be uncomfortable holding hands, introduce the pieces of rope and modify the instructions slightly.

Explain that the signs on the cards indicate the *kind of movement* required of each player's left hand. If a person is wearing a "+" sign, his or her left hand must move in the *same direction* (up or down) and *distance* (number of inches) as his or her right hand, after a 1-second delay. For those wearing a "-" sign, their left hand must move in the *opposite direction* and same distance as the right hand, after a 1-second delay.

Demonstrate. Point out that everyone is wearing a "+" sign, so their left hands must move in the same direction as their right. Send a practice "signal," or pulse, to the person on your left by raising your right hand 2″ above waist level and then, one second later, raising your left fist 2″. Point out that the person on your left, having felt and seen his or her right hand rise 2″, must now move his or her left fist up by the same distance.

Invite everyone in line to lower their hands and shake out their arms to relax. Ask if they have any questions. Then ask the observers to draw a rectangle for a behavior-over-time graph on their sheets of paper or in their notebooks. Explain that observers will be drawing one behavior-over-time graph for each trial. Each graph will have two curves. The first curve will show the position-over-time of your (the facilitator's) right hand. The second will illustrate the position-overtime of the ball held by the person at the other end of the line. To illustrate what you want, leave the line and use the chalkboard, flip chart, or white board to provide a sample illustration. For example, assume that the facilitator's right hand drops 2″. Assume further that there are ten people in the line, each pausing for one second between the movement of their right hand and the movement of their left. Then the behavior-over-time graph would look like this:

Height (from waist)

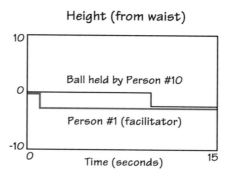

Now see what happens when the signal travels all the way down the line. Rejoin the line and hold your right hand in front of you, palm down. Form a fist with your left hand, and hold it in front of you. Ask the player to your left to place his/her right hand palm down on your left fist. Ask everyone else in line to get his/her hand in the proper position. Remind the observers to watch what happens next.

Announce that you are going to raise your right hand 1 inch. Ask the observers to draw the behavior-over-time graph that illustrates what will happen to the ball. After they have done this, raise your right hand, then after 1 second, raise your left fist by 1 inch. One-by-one, players should now raise their left hand 1 inch. Each movement of hands down the line should be separated by 1 second.

Ask observers to note the positions of everyone's hands
and of the ball at the end of the line. If all the players
moved correctly, the correct behavior-overtime graph
would look like this:

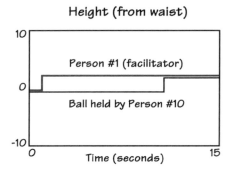

Height (from waist)

Person #1 (facilitator)

Ball held by Person #10

Time (seconds)

Now explain that you're going to lower your right hand by
2″. Lower your right hand, then your left hand, and ask
the rest of the players to follow suit. Ask observers to
note the positions of everyone's hands and the ball at
the end of the line over the time period of the exercise.
The correct behavior-over-time graph would be the one
shown initially above.

Now change the sign of one person in the line by flipping
his or her label card from "+" to "-" (see illustration,
above). Ask the observers to draw the behavior-overtime
graph that will result if you start by dropping your hand
2″. Pause while they do this. Now run the exercise again,
ensuring that the person wearing the "-" sign actually
reverses the direction of the signal. Have observers
compare their drawings with what actually happened. The
correct diagram is shown below:

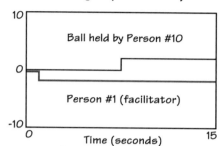

Run the exercise a few more times, each time starting out with a different signal. Before every trial, ask observers to draw what they think will happen. Then have them compare their predictions with what really happened.

Change a second person's sign to "-" so that now two people in line are wearing "-" signs. Ask observers to again predict the positions of players' hands and the ball over the time period of the exercise, if you start with, for example, a 1" increase. Point out that having two "-" signs in effect reproduces the original structure, in which all the signs were "+". That is, a set of links containing any even number of "-" signs (0, 2, 4, etc.) is positive overall. By positive, you mean that the final response is in the same direction as the initial input. Thus, for example, if you start things off by raising your right hand 2", the ball at the other end of the line will end up 2" higher. If you start by lowering your right hand 2", the ball will end up 2" lower. Demonstrate this. If observers and players are having trouble understanding that a chain of all "+" signs and a chain containing an even number of "-" signs create the same behavior, change a third "+" sign to a "-" sign and run the exercise again. Then change a fourth "+" sign to a "-" sign and run the exercise again.

Step 2: *Create a closed reinforcing loop.* Continue to ask that observers predict the behavior of each system you create.

Reset all the signs to "+". Then bend the line into a circle so that the person holding the ball has placed his or her left hand (still holding the ball) under your right hand,

which you're holding forward, palm down. Make sure that all the observers can see both your right hand and the ball.

Repeat the definition of "active" and "passive" hands and of "+" and "-" links.

Announce that you are going to raise your left hand just once by 2″, and then you will simply respond as indicated by your "+" sign. Ask what the ball will do. Give observers time to make their predictions. Then move your left hand up by 2″. Check to make sure that each successive person, starting with the player on your left, moves his/her fist in the proper way around the circle. Do not let people anticipate the motion; they should move their left fist only 1 second after they have actually felt their right hand move.

When the signal comes full circle, back to the ball and your right hand, pause one second. Then raise your left hand a further 2″ and watch that impulse move around the circle until it reaches you. Then pause one second and move your left hand 2″ higher. Let the signal travel all the way around the circle several times; eventually, a player will reach the limit of his or her ability to reach higher, and the signal will have no option but to stop.

Debrief
A Living Reinforcing Loop

NOTE: Conduct this debrief while still standing in the circle; you'll be trying and debriefing additional experiments as you go along.

Ask the players to release their hands, then invite them to describe what happened. Why was the behavior of the ball so much different this time? The answer, of course, is because you closed the loop for the first time, creating feedback. Since all the links were "+", the group created a reinforcing loop.

Point out that all reinforcing structures have limits that determine how far they can grow in one direction.

Test this idea by running the exercise again. Announce that the initial motion will be 2″ down. Give people an opportunity to predict what will happen. Most of them will correctly guess that the ball will move progressively down until someone's hands reach the floor. Then operate the loop by starting the signal yourself with your left fist. People will physically experience the idea that limits are inherent in reinforcing loops.

Step 3: *Create a closed balancing loop.*

Explain that the group will run another experiment. Say: "This time we'll introduce a single "-" link into our system. Most of you will still be representing a "+" link. You must still move left right fist in the same direction and distance as your right hand. One of you will represent a "-" link; you will move your left fist in the opposite direction and same distance as your right hand."

Tell that person to switch his or her sign from "+" to "-" (he or she will need to take off the card, reverse it, and put it back on).

Announce that the input will be 2″ down. Ask players to figure out how their own fist will move as the signal travels around the circle. Let players think about this silently for about 30 seconds. Then invite a few volunteers to demonstrate what they think their fist will do over the time period of the exercise.

Ask everyone to arrange their hands as before. Move your left hand down 2″. Watch to make sure that the "-" link behaves correctly and that the others are also correctly following the movement. This switch from "+" to "-" in one link causes people to move their hands down until the signal reaches the "-" link and then up afterwards, then down, and then up, as the signal passes around the circle and through the "-" link. Stop the group after several rounds, once everyone sees that the loop is oscillating indefinitely.

Debrief
A Living Balancing Loop

Invite the group to drop their hands and shake their arms to relax. Ask the following questions:

- "What happened? Why was the result so different?"

- "How long do you think this system could continue?" (The answer is "forever.")

- "What did it feel like to be in this balancing loop, compared to how it felt being in the reinforcing loop?"

- "When have you felt as if you were in a balancing structure in an everyday life situation?" (If participants are having trouble coming up with examples, offer the example of becoming alarmed about being overweight, losing weight, becoming complacent, eating more, gaining weight back, becoming concerned again, losing weight, becoming complacent, and so on. Because many people may feel sensitive about weight, frame the example as a story about yourself or about "all of us who have worried about our weight." Or offer the example of oscillation in an intimate relationship: You get frustrated by a conflict in the relationship, so you try hard to be nice and understanding. As a result, the relationship becomes peaceful. But then your frustration eases, and you feel comfortable voicing concerns again. Then conflict rises, frustration goes up, and after a while you try again to be nice.) Then ask players to try to generate more of their own examples.

Explain that the normal behavior of a balancing loop is oscillation; that is, the "motion" of the system continuously moves back and forth around a fixed point, just as people's hands were oscillating during the exercise.

Invite participants to imagine that they had been holding a pen in their left hand during the exercise, while a piece of paper moved past the pen slowly. Ask someone to draw on a flip chart what the resulting graph would look like.

Variations

Depending on the time you have, and on participants' enthusiasm, you can profitably design a sequence of revealing experiments. In each case, you'll ask one or more people in the loop to change their sign and then ask everyone to guess what the system's resulting behavior will be. You'll then conduct the exercise and compare the actual results with the predicted results.

NOTE: Watch carefully that people act out their sign ("+" or "-") correctly. It is very easy for someone who has just changed his or her sign to get confused or act as they did in the most recent exercise.

Play with these variations, but also try improvising your own.

Variation 1: With one "-" sign in the loop, an initial downward signal created an indefinite oscillation. Try making the initial signal an upward pulse.

Lesson: In a balancing loop, the direction of the initial signal makes no difference to the system's behavior. In a reinforcing loop, it does make a difference.

Variation 2: Ask the person who was the "-" link to become a "+" link. Then ask his or her neighbor to the right to switch from "+" to "-".

Lesson: The location of the negative link does not make a difference in this simple exercise. The crucial factor is the number of negative links.

Variation 3: Make two people "-" links and leave all the others as "+" links.

Lesson: Two, or any even number of, negative links act as a positive link.

Variation 4: Make one person a "-" link and all the others "+" links. Designate another player to move his or her left fist in the correct direction as the signal, but to double the distance. For example, if the signal the person received at his or her right hand was an upward 2″, the person would move his or her left fist up 4″.

Lesson: Exaggerating the response to the signal produces exploding oscillations.

Variation 5: Increase the time delays in the loop by asking people to wait 2 seconds between when they experience movement in their right hand and when they make the appropriate response with their left fist.

Lesson: The longer delay doesn't change the basic shape of the system's behavior, but it does stretch it out over time.

Variation 6: For a less informative—but more fun alternative—invite players to make sounds that go up and down in pitch according to whether they're a "+" or "-" link.

Debrief on Variations

Ask the following questions:

- "What general principle do these experiments illustrate?"
  Answer: A loop with an even number of "-" links (including zero such links) is reinforcing. A loop with an odd number of "-" links is balancing.

- "Does the location of the signs around the loop make any difference to whether the loop is reinforcing or balancing?"
  Answer: No.

- "Have you ever experienced (or can you imagine) the behavior of a system in which a link's polarity changed from '+' to '-'?" Give people time to reflect and answer.
  Example: the paradoxical situation in which people receive less money if they work more. This happens to some people who go from being on Welfare to getting a low-paying job. Many states suspend child-support or other kinds of payments when Welfare recipients get jobs—even if the wages are low.

- "In your own experience, which is more common—reinforcing structures with all '+' links or with some '-' links?"

- "How would you describe your feelings while you were part of the two different kinds of loops? In what other situations in your life have you noticed those same feelings?"

For example, one workshop participant responded: "On being a loop ... It is more anxiety producing to be a reinforcing loop, because you know the limit is coming. It is also more tiring. It is more rhythmic and natural to be a balancing loop. The up-down rhythm becomes like a pulse or a heartbeat—life sustaining as opposed to draining."

EXERCISE LINK

Using **Living Loops** before **Postcards** can make it easier for participants to link stories with real-world behavior.

Voices from the Field

John Shibley explains:

"We sometimes act as if teaching about systemic complexity is easy—we'll just teach people how to draw causal loops. I think the opposite is probably true—that we have just scratched the surface in finding ways to illustrate systemic complexity. As I've worked with Bunny Duhl and Howard Berins (who developed a version of this exercise with Steven Cabana and I), I've encountered this lesson over and over. Bunny and Howard are family therapists who use some of the same system ideas that we do. However, they've developed ways to get complex systems—in this case, families—to reveal their behavior in real time. I see their methods as powerful complements to system dynamics."

Gillian Martin Mehers, Training Director for LEAD/International, a leadership development program active in over 20 nations, says "We used **Living Loops** during a recent train-the-trainer workshop in Bali, Indonesia.

"Our first approach to 'pure' systems thinking in this session was rather academic. We conducted a lecture on systems vocabulary, in which we introduced causal loop diagrams and positive and negative feedback, etc. Then we presented various examples of CLDs, which we walked through.

"We spent several hours on this lecture. At some stages, people seemed to understand, yet they still had fundamental questions about signs, links, and other concepts that seemed counterintuitive to them. "In the end, the lecture was dense, thought provoking, and, for some people, frustrating. Systems thinking promised hidden power, but most people could not entirely grasp it.

"That evening, we organized an impromptu session on systems games to help those who had not entirely understood the lecture to practice more. We started this evening session with **Living Loops**, to help people understand the basics of positive and negative feedback and actually feel the loops. This was amazingly successful."

Carol Ann Zulauf, professor of organizational learning at Suffolk University, wondered:

"What would it be like to introduce a systems thinking experiential exercise into a former Soviet country where their main teaching methodology was lecture? I was slightly hesitant to try this, but I knew the risk would be worth it. We tried **Living Loops** with the college students from the European Humanities University in Minsk, Belarus—and they loved it! There was lot of laughter during the entire exercise, and the students kept referring to the exercise throughout the course. They told me they 'wanted to do more exercises like that!'"

Here a workshop participant shares her experience with the **Living Loops** exercise:

"This exercise is a very effective way to demonstrate how reinforcing and balancing loops work. The movement and 'acting out' help to solidify something that can seem very theoretical and abstract. Particularly effective to me was the experience of the reinforcing loop—feeling the urgency of going up, up, up—and then seeing that sometime, somewhere, a limit would be reached."

Joe Lauer of Seed Systems shares his experience using **Living Loops** to discuss issues related to sustainability with a large group of corporate managers:

"**Living Loops** works beautifully to introduce the concepts of systems relationships and to help people get the concepts in their bodies—literally!

"I have found it a quick and simple tool to boost understanding of how systems function, that the quality of a system depends on the quality of the relationships in that system, and that one element in the system can alter the entire outcome of that system. One part or one person can make a difference in even the most complex systems."

MENTAL MODELS | TEAM LEARNING | SYSTEMS THINKING | SHARED VISION | PERSONAL MASTERY

Harvest

> "You are informed
> about where you
> are coming from,
> not where you are going."
> A proverb of the Maasai tribe

What if the three Musketeers had used the following rallying cry? "All for one, and none for all!" Would that cry have inspired the sword-wielding heroes to the great feats of bravery and teamwork that helped them protect the queen from the evil Duke Leval? Probably not.

Yet "all for one, and none for all" is exactly what occurs when a system's structure encourages individuals to take action for their own benefit—but doesn't motivate them to see the value in collectively coordinated actions. Overcrowded swimming pools, over-fishing, and jammed shopping malls during the holidays are just a few examples of this behavior pattern. Indeed, this pattern is so pervasive, in so many areas of our lives, that it has a name: "tragedy of the commons." When a tragedy of the commons situation kicks in, people who are acting to advance their own well-being cause the collapse of the very environment on which that well-being depends.

Harvest provides a visible and enjoyable means of exploring the tragedy of the commons archetype. This exercise also helps participants explore the phenomenon of "worse before better"—a widely observed tendency in complex systems. In "worse before better," the actions required to produce fundamental, long-term solutions often make the situation seem worse in the short run. When politicians or economists persist in looking at only the short-term indicators of success as they select policies, the long-term results can prove tragic. This fact has been graphically illustrated in many sectors of society, but especially in our flagrant overuse of natural resources such as fishing grounds. In some areas, over-fishing has destroyed fish populations' ability to regenerate themselves.

To keep using a resource in the long term, we often have to accept a short-term reduction in what we harvest from that resource. And to implement sustainable-use policies, we must understand the system's long-term dynamics, value our long-term (not just our short-term) welfare, and trust each other to observe short-term constraints. **Harvest** gives groups the opportunity to practice all of these principles.

⇨ To illustrate the tragedy of the commons archetype

⇨ To enable participants to quickly experience a dynamic behavior that often takes decades to unfold in the real world

⇨ To provide an opportunity to practice communication and decision making in a complex system

⇨ To illustrate the problems that a "free rider" can cause for a group that is trying to negotiate a compromise to serve its long-term goals. A free rider is an actor who attempts to gain the long-term benefits of the group's policies without personally paying the short-term price required to implement those policies.

☆ Understanding of how quickly and unexpectedly a "commons" can collapse

☆ Insight about how teams can improve group discussion when they're attempting to define and solve problems

Something about the sound of coins jingling at the bottom of a coffee can attracts attention. **Harvest** starts out with that sound, and it's a great way to give participants a break from more traditional lectures and discussions.

This game also reveals what can happen when a select few dominate a system to the detriment of the collective good. If you find yourself working with a group dominated by a few individuals who insist on getting their way, or who push for short-term benefits at the expense of long-term gains, you may choose to interrupt the conversation and play this 15–30 minute game.

To Run This Exercise

2–6 teams, each comprised of 2–6 individuals

15–30 minutes

You'll need to select a space that accommodates two kinds of activity. First, you will introduce and facilitate the game to the whole group of participants. Then, you will lead participants through a debriefing conversation. It is most convenient to conduct both of these activities using a flip chart placed in front of a sufficient number of chairs to seat your entire audience.

Also, you'll need a room that lets people break into teams of 2–6 people. These small teams will need to sit or stand far enough apart so that they don't overhear one another's conversations.

One large coffee can or some other opaque, metal container that can hold 50 coins. The container must be large enough that you can reach into it to retrieve a small number of coins. Five rolls (200) of nickels or some other coins of about

---

the same diameter—3/4" or 2 cm. One container per team (paper coffee cup, a small basket, or anything equivalent), numbered sequentially on both sides with easily visible numerals (1, 2, 3, and so forth); 10 slips of paper or 3"by 5" index cards per team; 1 large flip-chart sheet showing the following four charts (in this order):

Chart #1: Game Title: **Harvest**

Chart #2: Rules of the Game: You are part of a team of people who fish for a living. Your team's goal is to maximize its assets by the end of the game. Each fish you catch is worth $.05.

The ocean can support a maximum of 50 fish. We start the game with between 25 and 50 fish in the ocean.

We will play for between 6 and 10 "years," making one round of decisions per year.

With each decision round, your team decides how many fish it will try to harvest that year. You indicate your desired harvest by writing the number on a slip of paper, putting the slip in your "ship" (the paper cup or other comparative container), and taking your ship to the game operator.

The operator will fill orders randomly. The fish you catch are returned to you in your ship. If your order exceeds the number of fish remaining in the ocean, you receive no fish that year.

After all orders are processed, and your team's ship is returned, the fish in the ocean will regenerate according to the curve shown on Chart #3.

Chart #3: Regenerating the Fish

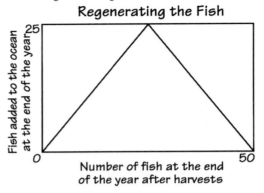

Regenerating the Fish

Fish added to the ocean at the end of the year (y-axis: 0 to 25)

Number of fish at the end of the year after harvests (x-axis: 0 to 50)

Chart #4: Steps of Play:

1. Decide on your team's long-term strategy.

2. With each decision round, select the number of fish you wish to harvest this year.

3. Record the number on a slip of paper, insert the paper in your ship, and take the ship to the game operator.

4. Harvest requests will be filled in random order.

5. Receive your ship, remove the fish, and start again with Step 1.

set-up Put 40 coins in the "ocean" (the coffee can). Put the remainder of the coins in a nearby container that is not accessible to participants. Put 10 slips of paper in each team's "ship" (paper cup). Divide players into roughly equal teams. Try for 2–6 teams, each comprising 2–6 members. Assign each team a number. Teams can sit or stand anywhere in the room. But, they should be far enough from one another that no team overhears another's strategy. They should also be close enough to the front of the room that they can see the charts and follow your instructions. If time allows, you may also encourage teams to give names to their ships.

Instructions

1. 2. 3. 4.

Step 1: Compose the teams that will play the game. Ask the members of each team to stand near their fellow members. Introduce the exercise with something like: "Congratulations! Each of you has just become a member of a fishing company. We start with a bountiful ocean." (Hold up the coffee can and shake the coins in it loudly.)

"Your team's goal is to maximize its assets by the end of the game. For this purpose, each team has a state-of-the-art fishing ship." (Hold up one of the paper cups.)

Now read aloud slowly through the rules on Chart #2. Answer any questions.

Explain the curve in Chart #3. "The curve means that if there are no fish left in the ocean after all orders have been filled, then no new fish will be added to the ocean. But if, for example, there are 25 fish left after all orders are filled, then 25 new fish will be added, to reach the ocean's carrying capacity of 50. If there are 38 fish remaining, 12 will be added."

"We will play 6 to 10 rounds. Each round represents one year."

The number of rounds you play will depend on the time you have available. Each round lasts approximately five minutes.

Step 2: Display Chart #4, Steps of Play. Leave this where it can be seen by all teams during the game. Give the teams a few minutes to discuss their long-term strategy and to submit their first fish request.

Step 3: Fill requests in random order. After gathering all the "ships," place them on the table in front of you, close your eyes, and mix the ships up. Open your eyes and arrange the ships in a straight line—left to right—visible to all participants. You do this mixing, because it is important that you fill orders in random order. Ship #1 should not necessarily be the first one to have its orders considered. Nor is the first team to hand in its ship guaranteed that they will have first call on the remaining fish.

Step 4: Pull the paper from the left-most ship. Do not reveal the size of the request. If there are enough coins in the "ocean" to fill the request, remove the requested number of coins from the can and put them in the ship. Then fill the orders from the next ship in the line, and so forth. If one order is larger than the number of fish remaining in the ocean, return that paper to the ship with no coins and go to the next ship. When you have processed all the orders, return the ships to their respective teams.

Step 5: Ask the teams to decide on their next order. While they're doing that, count the number of coins in the ocean and consult the Regeneration Curve to decide on the number of new fish to add to the ocean. This is quite simple. For any number of fish in the ocean between 25 and 50, you simply add enough coins to bring the total back up to 50. Below 25 coins, you add a number equal to the number remaining in the ocean after processing all the orders. For example, if there are 12 fish (or coins) left in the ocean, you would then add 12 more coins. You may either count the coins physically in the can or keep track, using a piece of paper, of the initial number minus the total of what you put in the ships.

Step 6: Collect the ships for Year 2, process the orders, and continue. If the teams quickly catch all the fish, let them go through one or two more yearly cycles experiencing the consequences of their mistake—no catch. Then stop the game. If you can see that the entire group has adopted a strategy that will keep the fish population sustained around the point of maximum regeneration, you can also stop the game. But with most groups, you will have to go through at least 6–8 cycles before participants experience the consequences of their decisions.

Debrief

Typically one or two teams will pursue an aggressive strategy and place large orders early in the game. That causes the fish population to decline, pulling down the possible harvest for everyone. Sometimes there will be a serious effort to coordinate all the teams' decisions and produce a total harvest that can be sustained over the entire period of the game. But that effort usually fails. Either it is ignored by one or two teams or it is based on a false estimate of the maximum number of fish that can be harvested annually.

The regeneration curve, Chart #3, shows that 25 is the maximum number of fish which can be added to the ocean each year. Therefore 25 fish/year is the maximum number that can be harvested sustainably. Over 10 years, 250 could theoretically be harvested without reducing the fertility of the ocean. Divide that number by the number of teams, multiply by the value of each fish, and you have the maximum average wealth possible per team. If any team fails to reach that level of assets, it is typically because there was over-harvesting early in the game.

Have each team report its wealth on the flip chart in front of the room.

Then lead the participants through a discussion about their experience.

- What happened in this game?

- Who was responsible for this result? Actually, in **Harvest**, the structure of the game bears more responsibility for the collapse in the fishery than any individual does.

- What would have been the maximum possible wealth available to all the teams in this exercise?

- What wealth did teams actually achieve?

- Was there a winner in the game?

- What policies would you have to follow to achieve maximum wealth for all the teams? Why might these policies not be followed?

- Where do you see examples in real life of the behavior we witnessed in this game?

- What policies could be followed in real life to produce a more sustainable result?

Ask participants to draw the behavior-over-time graph for their total harvest during the game. Then show them the following causal loop diagram with a balancing loop involving Actual Harvest and a regeneration loop that is initially balancing. But it becomes reinforcing, in a downward direction, once Fish Population declines below 25. Ask them to use this diagram to explain the results they received in the game. They could use it also to identify possible new policies governing Desired Harvest.

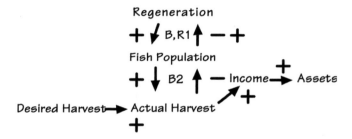

resources A more elaborate version of this game is available for those who have the time and the goals to warrant its use. *FishBanks, Ltd.* is a computer-assisted, role-playing game for groups up to 50. It takes two to three hours to play, but it is rich in learnings. Contact: Sustainability Institute 3 Linden Road Hartland VT 05048, www.sustainabilityinstitute.org, 802-436-1277

For an excellent discussion of the tragedy of the commons archetype, see *Systems Archetypes II: Using Systems Archetypes to Take Effective Action*, by Daniel H. Kim. Pegasus Communications.: 1994).

Triangles

"[Organizational] challenges are predictable. They arise as natural counter pressures to generating change, just as the need for soil, sunlight, and water arise as natural limits when plants start to grow. Though they often appear as seemingly independent events, they are interconnected and interdependent. There are high-leverage strategies that can help teams and individuals deal with each challenge separately. But the greatest leverage comes from understanding them as an ensemble of forces."
Peter Senge
The Dance of Change

At the core of systems thinking and system dynamics lies the premise that the structure of a system drives the behaviors we observe. By structure, we mean all of the elements or agents that are part of a system and the manner in which a system's elements are organized or interrelated. The structure of an organization, for example, could include the physical lay-out of the offices, the roles and responsibilities of employees, the incentive systems and information flows.

In **Triangles**, a group of participants creates a simple dynamic system. Through their own movements and interactions, group members experience several important concepts related to causality, such as interdependence (mutual influence among all the system's parts), feedback (the return of information within the system), the tendency for systems to stabilize, delays (time lag between actions and effects), the impact of a system's structure on its behavior, and leverage points (points where changes to the system can produce significant changes in its behavior). This simple, easy-to-run exercise is short on set-up and long on experience-based learning.

⇨ To illustrate, via actual experience, several key characteristics of dynamic feedback systems such as interdependence, balancing processes, delays, the relationship between structure and behavior, and leverage points

⇨ To give participants practice using "influence" diagrams to describe observed and/or experienced dynamics

⇨ To animate the concept of an exogenous input (an outside influence on the behavior of the system)

☆ An opportunity to deepen one's understanding of the relationship between structure (illustrated in a causal diagram) and systemic behavior through physical experience of the concepts

☆ Illustration of consequences that stem from manipulating high-leverage and low-leverage points in the system

In his best-selling book, *The Fifth Discipline: The Art & Practice of the Learning Organization*, Peter Senge offered this simple, yet profound axiom: "Small changes can produce big results—but the areas of highest leverage are often the least obvious." (p. 63) Here, he refers to what systems thinkers call "leverage points," well-timed, well-placed actions that can produce significant, lasting improvements. Most people immediately grasp the concept of leverage points, but to spot them in an actual system is often more difficult. This exercise quickly illustrates the concept of leverage points through concrete changes made to the group's structure.

Because most causal-loop diagrams depict a variety of intangible influences, viewers often do not fully understand what the diagrams portray. Thus they struggle to relate the system's structure to its different behavior patterns. **Triangles** lets you create a living structure in which the only influence is physical proximity—a concept that all participants can understand. We appreciate this exercise for the ease with which it conveys a seemingly abstract concept: the structure of a system—the relationships among its parts—produces its behavior. A change in the structure of a system typically produces different behavior. In **Triangles**, the structure of the system is created by the arrangement of the participants in the exercise.

To Run This Exercise

 Ideally, 10–50; if your group exceeds 50, split it into two groups with less than 50 people

 20–30 minutes

 A square or rectangular area 20'–30' long on each side, free of obstacles, and large enough for all participants to move around freely

 A large flip chart or chalkboard, several flip-chart markers or pieces of chalk. A set of numbered self-sticking tags—one tag for each participant beginning with 1 and continuing (2, 3, 4, etc.)

 On the flip chart or chalk board, write down in the shape of a circle that has a diameter of about 2', a series of equally spaced, sequential numbers, starting with 1. The numbers must go up to the total number of participants in the exercise.

Instructions

Step 1: *(optional)* Ask participants to think of a time when they planned for something to happen but then got a very different result. Ask one or two people to briefly share their stories. Then say: "Through this exercise, we can gain insight into how systems can make things turn out differently than we expected."

Step 2: Ask the group to form a circle in the middle of the room, facing the center and standing about 3′ or 4′ apart if space allows. Explain that they will spend the next 10–15 minutes forming a system with a simple cause-and-effect structure. Distribute the tags to participants, making sure they understand that the numbers have no special significance. Participants stick the tags on their shirt fronts.

Step 3: Say: "I will ask each of you to select two other people in the circle. These will be your references. But in picking your reference, please observe two conditions.

1) All odd numbered participants, 1, 3, 5, etc., should choose person number 2 as one of their references.

2) No one should pick as a reference any person who has _____. Name some easily visible characteristic that is shared by only a few people in the circle; for example, 'a green shirt, a yellow belt, or sandals.' Of course the feature you designate must not be displayed by person number 2.

Pick your two references, then memorize your own number and those of your two references."

Caution

Be careful to find some basis for exclusion that is not likely to embarrass the people who exhibit it. Items of clothing are generally acceptable. Physical features are better avoided.

Step 4: Say: "When I say 'Go!' your goal is to move around the room until you are equally distant or equidistant from your two references. You can be very near or far from each of them; that makes no difference. But you have to be the same distance from each. I may intervene at some point and steer or stop one of you gently by placing my hands on your shoulders. If I say 'All stop!' then everyone should halt immediately and remain standing in place. Any questions?"

To show what you mean, invite two participants to join you in the middle of the circle. Tell them to stand in two specific locations and then illustrate how you would have to move to make yourself equidistant from them. Illustrate being close and equidistant and far away and equidistant. When you are equally distant from both of them, ask one to move several feet and then show how you would have to move to maintain equidistance. Again, ask if there are any questions.

Step 5: *Return your two demonstrators to their original positions, and get yourself out of the circle.* Say: "In a moment, I will ask participants 1, 2, and 3 to get equidistant from their references. Everyone with a number of 4 or higher must not move. What will happen, when I say, 'Go!'?" Let participants briefly discuss possible answers to this question. Then make the announcement: "Remember: only the first three people can move. . . . Go!"

Watch what happens. Typically, the three people move briefly, and then the system comes to rest.

Step 6: *Return participants 1, 2, and 3 to their original locations.* Say: "In a minute, I'm going to ask everyone to get equidistant from their references. What will happen?" Let people speculate about this, then say, "Go!"

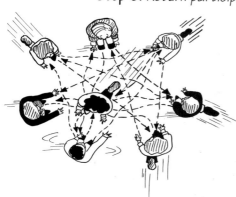

Watch what happens. Most often, people will mill around the room for a minute or two, before everyone slowly comes to a stop. Be patient. It may take longer to reach equilibrium, but eventually the participants will come to rest.

Step 7: *Illustrating high and low leverage.* Ask everyone to return to the original circle. Point to the people who have the characteristics that you defined earlier to exclude candidates as references. For example, the four people who have a green shirt.

Explain that in the next part of exercise, only those four people must quit moving when you say, "Stop!" All others will continue to move until they are equidistant from their references. Ask the participants how their system's behavior will differ from the last exercise. Give them time to reflect and respond. (In fact it should not differ, because the people you are stopping do not exert any influence on others in the group. But do not announce this answer; let the participants discover it through their own activity.)

Tell the group to "Go!" After three seconds in which everyone moves around trying to get equidistant from his/her references, say "Stop!" The previously designated people should stop immediately. It will take some time for the rest of the group to come to a halt.

Return to the original circle. Ask what happened and why. Now pick another group of four people, including participant number 2. Tell this group that only they must stop moving when you say the word, "Stop!" Ask what will happen. Then say, "Go!" and after three seconds, "Stop!" The four people you designated should stop immediately. The rest of the group typically will also stop very quickly. Return to the original circle and discuss what happened. Person number 2 is a "high leverage" point in the system, since half the participants are using him or her as a reference. Stopping person 2 through an exogenous influence has major influence on the whole system. You have illustrated the concept of high-and low-leverage.

Step 8: Depending on the time you have and your goals, you can conduct other experiments. In each case, make sure that everyone understands what you are going to do. Ask them ahead of time, "How will this change influence the behavior of the system? Why?" Give them time to reflect and to answer your questions.

Announce that you will stop a randomly selected person while all the members are trying to get themselves equidistant from their reference.

Announce that when someone sees their references move, they should count to three slowly before going to the spot that would place them equidistant from their references. That is, determine the location of your two references, identify the spot that would make you equidistant from them at this moment. Then count to three and go to that spot. Once you are in that spot, see where your two references are, and determine where you should position yourself to be equidistant from them. Then count to three slowly and go to that spot. Continue until your references quit moving.

Then, announce that everyone should stay at least five feet away from their references at all time while they are striving to become equidistant from them.

Then, announce that everyone should stay no more than two feet away from their references the entire time they are striving to become equidistant from them.

When the group has come to rest, announce that you will move person number 2 ten feet from her current position and hold her there. Announce that you will move one of the people excluded as a reference ten feet from his current position and hold him there.

Step 9: Construct the influence diagram for this system. Ask each person to go up one at a time to the circle of numbers that you drew on the flip chart or chalkboard. Each participant draws two lines, each line extending from their own number to the number of one of their two references. This is not, strictly speaking, a causal-loop diagram. But it does show influence. Moreover, you can use it to identify the high-influence and low-influence individuals, person number 2, and the people who were excluded. Any number that has only two lines connecting it to other numbers represents a low-influence individual. Any number that has more than two lines connecting to it other numbers represents a high-influence individual. Moving low-influence people in a group will produce little movement as you run the exercise; moving high-influence people will immediately produce a great deal of movement. If time permits, experiment with moving some other high influence individual, other than person number 2.

Variation

Ask each person to identify just one reference. Explain that when you say "Go!" everyone must try to get five feet from his/her reference. Ask what kinds of behavior people expect to see. Give them time for speculation, then perform the exercise.

Debrief

Ask group members to take their seats and share their general impressions, feelings, and observations about the exercise. Many insights should already have emerged during the conversations after each experiment. Now it is time to summarize. Some possible questions to ask:

What principles or general rules would help you explain and predict the behavior of the system to a newcomer entering it for the first time? What real-life behaviors does this exercise remind you of? Where have you seen high- and low-influence individuals or policies in your own organization? If you were trying to promote stability or change in your own organization, what general principles in this exercise would help you attain your goal?

When we ran this workshop recently, participants made the following observations:

The greater the number of components (in this case, people) in a system, the more difficult it is to change the system's behavior. The fewer the components or elements in a system, the more chaotic it becomes and the easier it is to influence and apply change.

Caution

If you are conducting this exercise with young adults or teenagers, you may want to be sensitive to the language you use. For example, use the term "low-leverage" rather than "low-influence" individuals. If you are sufficiently well informed, you may include less popular individuals as the references.

Voices from the Field

Gillian Martin-Mehers, Training Director for LEAD International, has this to say about **Triangles**:

"We played **Triangles** with an international group in Thailand. It was very successful in helping people understand the concept of leverage points in general. We needed to be careful about our terminology and not have people feel that if they did not "influence" people, they were not important or popular. In other words, non-influence was only in terms of the system and not in general! It might have been good to precede this exercise with a discussion or presentation introducing the concept of leverage points. We just played it. Though people understood it in general, it might have been more meaningful for them to have had a quick oral introduction of leverage points before playing."

Nan Gill from the Waters Foundation has used **Triangles** with students and educators numerous times. Here is a description of her experience:

"I have used this activity many times in school settings to help students and staff members better understand fundamental systems concepts. After experimenting with many introductory activities, I have found this to be the most powerful in building understanding that can be drawn upon for future learning. Regardless of age, participants always notice and describe the concepts of behavior over time, interdependence, delays, and how structure generates behavior. The beauty of this—and the power—comes from the fact that they use many different words and analogies to do so. Putting things in their own words strengthens their own learning and expands the learning of others. The physical acting out of this system, and then discussing what happened, supports people in seeing and understanding these concepts in far more meaningful ways than I could ever hope to simulate through verbal explanations alone. And what's more, everyone has a lot of fun!"

To see a variety of systems games geared toward students and educators, see the Waters Foundation website: www.watersfoundation.org.

resources For a discussion of the relationship between a system's structure and its behavior, see

"Lowering the Water Line: Seeing How a System's Structure Influences Behavior" (p. 31–31), in *When a Butterfly Sneezes: A Guide for Helping Kids Explore Interconnections in Our World Through Favorite Stories,* Volume I in the *Systems Thinking for Kids, Big and Small Series.* Waltham, MA: Pegasus Communications, 2001.

Avalanche

"A ferry boat was once crossing a river. Suddenly the boat struck a rock, and water relentlessly poured into the cabin. The passengers were frightened out of their wits. Only one man sat calmly as if nothing had happened and even laughed at the way the others were so alarmed. 'Don't worry! It's not our problem,' the man said. 'It doesn't matter if it's leaking, because it's not our boat.'"

Traditional Chinese joke

What makes traffic jam? How do epidemics spread? Why does the stock market fluctuate?

What do these questions have in common? They all attempt to understand emergent behavior, that is, behavior that arises from interconnections among elements and/or agents in natural and social systems. These behaviors are distinctive and often confounding.

One way to answer these questions is by focusing on self-organizing behavior, i.e., the patterns of behavior by the whole system that emerge from the interactions among large numbers of its components each following simple rules. For example, to understand how a bird flock keeps its movements so graceful and synchronized, one approach would be to simulate (using a computer) the thousands of interactions among the birds. In the case of a flock of birds, there is no lead bird in the flock that dictates the movements of all the others. Rather the flock can form as a result of the independent interactions of all of the birds. In fact, the birds might not even have the goal of making a flock, but each bird is really just trying to stay near (but not too near) some other birds. Studying systems by investigating the rules followed by the individuals in the system (whether it be people in a traffic jam or migrating turtles), is an approach used by the complexity sciences.[1]

Another way to answer the above questions is through system dynamics, a method that uses causal diagrams representing stocks and flows. While decentralized thinking focuses on the simple rules followed by individual agents, system dynamics focuses on (and attempts to model) the aggregate behavior of a population as it is influenced by particular policies or norms.

Which approach is better? In our view, the two perspectives are complementary. In this game, we take you through two debriefs in an attempt to raise our collective awareness about the different ways to look at complex system behaviors. One debrief is appropriate for exploring dynamic behaviors from a systems thinking/ system dynamics perspective, and the other explores issues related to self-organization and decentralized thinking.

⇨ To explore the concept of emergent behavior

⇨ To explore the dynamics of self-organization

⇨ To let participants experience physically the dynamic of "escalation"

⇨ To give participants practice with organizational learning skills

⇨ To create an opportunity for discussing the costs and benefits of breaking the rules

[1] Our appreciation goes to Eric Klopfer for his careful review of this exercise.

☆ A shared experience that can provide a rich metaphor during discussions about the conflict between individual norms and group goals

☆ Appreciation of the need for balance between action and reflection in carrying out a task

☆ Awareness of the counterintuitive nature of complex systems (when a system has many interconnections, it becomes difficult to anticipate the consequences of a decision or event)

CONTEXT

Many games are geared toward helping members of a team recognize that individual success depends on coordination among all team members. But **Avalanche** achieves that purpose quickly and with very surprising results. The key lesson behind **Avalanche** is that a set of individual rules for behavior ("Don't let your finger lose contact with the pole") can produce a result totally opposite from what the group wants to achieve.

You will see that there are actually two sets of rules being followed in this exercise— the ones that you as the facilitator define, i.e., "don't let your finger lose contact with the pole," and the individual rules that people must actually follow if they manage to coordinate their efforts and lower the pole.

Indeed, individuals will fail to attain their goal, until they quit focusing on their own role and work together to understand the behavior of the overall system. Teams are challenged in this exercise to devise effective plans and perhaps define different roles for their members, to prevent the actions from being really decentralized. To accomplish the main task at hand, teams may even need to break a rule or two.

The organization Outward Bound has taken thousands of participants into the wild for self-discovery and team building. The time spent around a fire in the evening on these trips gives participants great opportunities to develop new team-building exercises. During one of these evenings, someone picked up a tent pole and created the prototype of **Avalanche**. Anonymous facilitators have adapted and refined it, and it finally came to us via Professor Michael Gass, senior faculty member in the Outdoor Education Program of the University of New Hampshire. Mike is a brilliant innovator, facilitator, and trainer who uses extensive therapeutic programs with family groups. He calls this exercise "Lowering the Bar" or "Helium Pole." We always prefer names that don't give participants clues to a game's action or lessons, so we've changed Gass's designation to **Avalanche**. The name references the fact that avalanche probe rods are often used to conduct the game. Despite the name change, the exercise remains in essentially the same form developed by Gass. We have added features to make it more reliable and have sketched out a set of debrief questions that are designed to help participants explore several systems thinking-related ideas

To Run This Exercise

 Teams of 8–15 individuals

 10 to 50 minutes

 Enough room that the members of each team can line up shoulder to shoulder, work with their pole, and discuss the results.

 One light-weight, but rigid, pole per team; this can be a long wooden dowel of at least 1/4" diameter, an avalanche probe, 1/2" plastic plumbing pipe, or segmented tent pole; for each pole, 2 large washers with a hole diameter twice as large as that of the pole, and 2 washers with a hole diameter just slightly larger than that of the pole; duct, masking, or Scotch tape. One variant of this exercise also requires a small paper strip, about 1"x 3" for each participant.

 Assemble audience members into teams of 8–15. Then prepare one pole for each team. Each pole needs to be long enough to span the entire length of a team after its members have assembled themselves like a zipper— two

rows facing each other. Each row consists of team members standing shoulder to shoulder with fingers extended at waist height to make a cradle. The team's pole should be at least the same number of feet long as the number of people on the team.

To prepare the poles, reconfigure a segmented tent pole into longer or shorter lengths to accommodate the number of people on a team. You can also cut down half-inch plastic plumbing pipe or temporarily assemble it into longer pieces using standard plastic sleeves.

Once you've configured your poles, wind a piece of tape around the pole about 1″ in from the ends of every pole. Use enough tape to form a small ridge about 1/8″ high.

Distribute one pole to each team. Assign two assistants to each team. The assistants could be chosen from among the participants, if necessary. The assistants' role is to line up the members of each team, put 2 washers on the each pole, 1 between each pole end and the tape ridge. Then lay the pole across players' outstretched fingers when you give the command, and ensure that the game's rules get followed.

Instructions

Step 1: *Assemble teams.* The members of each team arrange themselves in two lines of approximately equal length, shoulder to shoulder, facing each other as shown in the diagram above. Each person holds out one forefinger at about waist height, pointing to the other line. All palms should be facing down.

Step 2: *Explain the exercise:*

- "One of your assistants is going to rest a long pole between the two lines across the tops of your fingers. (Make sure your fingers are all at the same height.) On each end of the pole, a washer will be hanging loosely."

- "Your team's goal is to lower the pole smoothly to the floor, rest it there, and remove your fingers—all while making sure the washers stay on the pole ends. There are three rules:

  Rule #1: You cannot hold onto the pole; you can only support it with the top of your finger.

Rule #2: You absolutely must not lose contact with the pole at any time. If your finger does lose contact with the pole, raise your free hand. The assistants will seize the pole, and your group will start over.

Rule #3: If one or both washers fall off the pole, the assistants will seize the pole, and your group will start over."

- "Imagine that the pole is a complex task that your group needs to accomplish. One washer represents your budget constraints; the other, your quality standards. You need to finish the job without 'dropping' either."

- "Any questions?"

Step 3: Try it! Ask the assistants to put the washers on the pole ends on the outside of the tape ridges. Ask assistants to lay the poles along the teams' fingers. Say "Go!" Typically, the poles will immediately start to rise as each person tries to follow the rule about keeping his or her finger in contact with it. Often, the poles will rise until they reach the limit of the shortest person's ability to reach up. Most participants will be surprised or even embarrassed by this turn of events. Some of them may even accuse you of playing some sort of trick.

Depending on the time you have and your goals, you may remain silent, or have your assistants seize the pole if any of the rules are broken. If you seize the pole, you can let the participants discuss their experience, make a new plan, and try again. Or you can go straight into the debrief.

If you *do* let players try again, emphasize Rule #3 ("If washers fall off the pole ends, the team has to start over"). Three to six trials should suffice for the team to master the task. If players are having real difficulty keeping the washers on, consider starting over but using the smaller washers. If they perform the exercise on the first try without the pole rising, try the following variation.

Variation

To reinforce the rule that no finger must lose contact with the pole, we sometimes give each person a small strip of paper (about 1"x 3"). Team members hold the strip in their left hand and extend their right forefinger to receive the pole. They lay one end of the strip on top of their right forefinger where the pole will rest on it. As the pole comes down on their finger, pinching the paper, they release their hold on the paper with their left hand. During the exercise, if the paper strip flutters to the floor, it reveals that their forefinger lost contact with the pole.

You may try the simpler variant first, but have the slips ready in case the teams do not produce the desired, counterintuitive rise in their poles.

Debrief
Systems Thinking and Team-Learning Lessons

In **Avalanche,** participants see that a set of individual rules for behavior ("Don't let your finger lose contact with the pole") can produce a result totally opposite from what a group intends. Indeed, this result will keep happening until people get out of their own role and work together to understand the behavior of the overall system. Then teams will need to devise effective plans and perhaps define different roles for different members. To accomplish the main task at hand (lowering the pole without dropping the washers), teams may even need to break the rule about not letting their fingers lose contact with the pole.

An interesting way to make this point is to say, "You were all doing this" (put your hand out with forefinger extended and palm down, just as the participants did) "but you were acting like this" (put both your hands up on either side of your head, like a horse's blinders).

To draw the systems thinking and learning lessons out of this exercise, ask these questions:

- Who caused the pole to rise?
  Answer: No one caused it; the rules of the game produced this behavior.

- Where have you seen examples of this sort of behavior in your own organization?

- What does it take for you to start thinking about the whole system instead of just your own effortsto follow the rules?

- What eventually stopped the pole from rising the first time? (Typically, it's the limit to which the shortest person could reach.) What limits exist in your own organization to stop you from moving in the wrong direction?

- What was the difference between the first (unsuccessful) and the last (hopefully, very successful) trial?

- What had to happen for your group to work together effectively?

- What conditions could you create in the future that would make it easier for your team to learn a new approach more quickly?

Decentralized Thinking Debrief

In **Avalanche**, initially no one person is in charge, and the individuals are essentially following individual rules ("Don't let your finger lose contact with the pole"). In most cases, each person keeps his or her finger on the pole until the pole goes up so far that one person can't reach any higher. Each person is trying to follow the "rules," as you have defined them, but each person controls their own behaviors and the result is the pole rises.

We can think of the group as a decentralized system. In a decentralized system, no one part of the system is "in charge." Behavior patterns emerge from interactions among individual agents. Examples of decentralized systems are all around us and can be found in both natural and human systems, for example, birds in a flock, the human immune system, ant colonies, cars on a highway, and market economies. Through self-organization, individual agents (such as, birds, molecules, ants, cars or people) follow a simple set of rules. Through interactions among these individuals (which involves a process of feedback and disequilibrium), a pattern eventually emerges, that is the pole rises. In the **Avalanche** exercise, the high-order form shows up in the group's ability to coordinate their actions and to lay the pole, smoothly, onto the ground. Only when the system becomes directed does the pole get to the ground.

To draw out lessons about decentralized thinking, share the above information and examples with the audience. Ask these questions:

- If a centralized system is characterized by centralized control (for example, a hierarchical organization with a dominant leader at the top), and decentralized systems are characterized by individual agents following simple rules and interactions, how would you describe **Avalanche**?

It is an exercise where, finally, a dominant leader gives everyone simple rules to follow individually. This shows the transition from decentralized to centralized systems. It should be clear that decentralized systems don't always have the opposite behavior that the individuals want (e.g. bird flocks). Said another way, decentralized systems are not always chaotic and selfish.

- What strategies did you use in tackling this exercise?

- What would a decentralized versus centralized strategy look like in the case of **Avalanche**? What did it take for a desirable outcome to emerge?

- What strategies can we use to think about, understand, and intervene in decentralized systems?

Ethics Exploration

This exercise raises another set of issues related to the ethics of following and breaking rules. It is virtually certain that someone will lose contact with the pole and not announce it by raising a hand, thus breaking a rule. If appropriate, you can help the group explore the issue of norms violation by asking these questions:

- What was the rule related to your fingers and contact with the pole?

- What was supposed to happen, if you broke this rule?

- Did anyone lose contact with the pole and not raise their hand?

- If so, what did you think to yourself to justify this behavior?

- More generally, when do you feel justified in breaking the rules governing a process?

- What would happen if everyone started breaking the rules whenever they thought it was justified?

- Did you consider stopping the exercise and discussing the possibility of breaking or changing the rules with all your team members?

- What would have had to happen for you to have done that?

- How do your thoughts about these questions relate to what you see going on in your own organization?

Voices from the Field

Our experience mirrors that reported by Michael Gass:

"I have used this activity with groups that are in strong systemic relationship, primarily with families and corporate groups.... This exercise is one of the best I have seen to illustrate [Peter Senge's] Escalation archetype. Other corporate archetypes abound within the experience, depending on the group's particular focus.... The exercise provides a rich medium for family members to project their relationships onto the experience. Positive elements of cooperation— family problem-solving and interactive dialogue—tend to occur, along with negative components of blaming, scape goating, and frustration.

"I have also used it with middle-school and university students to provide quick insight into the concept of inter-relatedness. With these groups, the experience seems to provide great new perspectives into the balance and interconnectedness of individual and group needs."

From Michael Gass, Lowering the Bar or Helium Pole, Zip Lines, Summer 1999, p 26.

Dr. Natalia Tarasova is Chair, Problems of Sustainable Development, at the Medeleev University for Chemical Technology in Moscow. She reports on using **Avalanche** with her students: "I told the students that the height of the bar was their level of resource and energy consumption. They needed to lower it greatly. Of course the bar initially went up. That led to many useful conversations about the rules in society that cause consumption to keep rising."

 Michael Gass, *The Book of Metaphors, vol. 3*, 1999. Kendall/Hunt.

V. Stephens Colella, E. Klopfer, M. Resnick (2001). *Adventures in StarLogo*. New York, Teachers College Press.

Mitchel Resnick. (1995). *Termites, Turtles and Traffic Jam: Explorations in Massively Parallel Microworlds*. Cambridge: The MIT Press.

Ralph Stacey, (2001) *Complex Responsive Processes in Organizations: Learning and Knowledge Creation in Complexity and Emergence in Organizations*. Routledge.

For a description of decentralized thinking as it being used by high school students and teachers, see *Adventures in Modeling: Exploring Complex, Dynamic Systems with StarLogo* (2001): Teachers' College Press by Vanessa Colella, Eric Klopfer, and Mitchel Resnick.

resources

Avalanche

MENTAL MODELS

TEAM LEARNING

SYSTEMS THINKING

SHARED VISION

PERSONAL MASTERY

Space for Living

"When several people's fingers burn in the fire, you rescue yours first."

A proverb of the Maasai tribe

Many well-intentioned efforts by a group to develop new and systemic insights into its organization's challenges founder on three deeply ingrained ways of thinking. First, team members tend to avoid pursuing lines of thought that might force the sharing of resources—they believe that "If he gets more, then I'll have less." Second, many people prefer to ignore a problem until it is widely perceived— "When I see it, I'll do something about it." Third, people often rely on tried-and-true methods— "If it worked before, it will work again."

This exercise can lead a group through a shared experience in which they confront the consequences of these ways of thinking. **Space for Living** has been adapted from a classic adventure-education game that is typically called Star Wars. We have changed the name and added some new rules to the classic version, so that the activity generates useful insights for adult participants in our workshops. This exercise also serves as an effective icebreaker.

⇨ To explore the phenomena of growth that reaches a limit

⇨ To demonstrate the need to be open to new plans or policies even while current ones seem successful

⇨ To illustrate some basic principles governing innovation and opinion change in groups

⇨ To provide a metaphor that is relevant to situations that arise when resources become inadequate to support the habitual way of doing business

☆ A shared metaphor that participants can use to address problems resulting from growth that reaches a limit

☆ A stronger sense of solidarity and familiarity among the members of a group

☆ A more realistic understanding of one's own reaction to scarcity, to atypical ideas, and to planning

This initiative may be seen as too childish to be respected by adult participants. But if you take the activity seriously and lay out the rules clearly and succinctly, **Space for Living** gives everyone a fun and profitable experience. We have seen some participants gain deep and lasting insights from the game.

This activity is also useful when a workshop session addresses problems of resource problems, i.e., limits and scarcity. We usually do not use it at the beginning of a workshop. We find it is best to wait until participants have gotten to know each other a bit before asking to put themselves in close physical proximity.

To Run This Exercise

A minimum of 10–15 participants is required; ideally, 25. With groups larger than 30, split into subgroups of 15–20.

15–30 minutes

Outdoors is best if the surroundings and climate aren't too distracting. If you use a space indoors, an open floor area of at least 20' by 20', clear of any objects, for a group of 15–20 people

 In total, about 6' of rope per participant (cotton clothesline works well). Initially you will cut this up into pieces of different length and tie each piece into a loop. In subsequent sessions of the game, with other groups, you can reuse the rope by tying several pieces together to get the correct number of loops of the proper size.

 Use 60% of the lengths of rope to make loops large enough to encircle one pair of feet with no part of the shoes touching the rope. Each loop will require a piece of rope about 3 1/2' long. Form loops by knotting the ends of the rope lengths together.

Use 20% of the rope lengths to make loops large enough to encircle the feet of 2–3 participants. Each of these loops will require a piece of rope about 7' long.

Use all but one of the remaining rope lengths to make loops big enough to encircle the feet of 5 participants— about 11' each.

Make one loop big enough for about two-thirds of the participants to stand within it very closely packed— this may be from a piece up to 20' longer or more, depending on the group size.

The total number of loops you will make should be equal to the number of participants.

For Variations 1 and 3 (see instructions below): Put all the loops on the ground so that they're at least one foot away from each other. Pull each loop into the shape of a circle. Do this in a way that leaves clear ground where everyone can stand while you are introducing the game.

For Variation 2 (see instructions below): Follow the same set-up instructions for Variation 1, except use one rope length to form a line on the ground between all the participants and the rope loops.

Instructions

Variation 1

Step 1: *Gather your group near the rope loops.* You will need at least one supervisor for each subgroup. Ask all participants to stand inside a loop so that their feet don't touch the rope.

Step 2: *Explain the game's rules:*

- "Imagine that the space you're standing in represents an important resource. For you to 'survive' to the end of the game, each of you needs to keep finding space within a loop—it is essential, in each round, that you find a place where your feet touch the ground inside a loop. Until I see that everyone has the resources they need to survive, I won't be able to move to the next round of the game."

- "When I see that all of you have 'space for living,' I will say 'Switch!' Then, if possible, you must leave the loop where you are standing and find space for your feet within a different loop. Again I will wait until everyone has found the space they need, with their feet not touching the rope or the ground outside the rope. Then I'll say 'Switch!' again."

Step 3: Look around to ensure that everyone is standing inside a loop with no illegal touches. Then say "Okay, Switch!" Wait for all participants to move to a different loop and position themselves inside it, without any foot touching the ground outside a loop.

Step 4: Repeat the exercise one more time. However, just as you say "Switch!" pick up several smaller loops as soon as they're empty. It helps to have a colleague assist you with this. [If someone refuses to vacate the loop you wish to pick up, just untie the knot and remove it from around him or her.]

There may be brief panic, until participants realize that more than one person can stand within the larger loops. Observe how this realization first occurs to an individual, and then moves through the group.

Step 5: Continue with several more rounds, each time removing several smaller loops.

Step 6: When only 1 or 2 loops remain, it will be impossible for all the participants to stand completely within the space the loops provide. At this point, some participants may give up and stand by the side. Or some may start creating human pyramids—for example, by trying to carry colleagues on their shoulders. Don't permit either of these strategies. The first response violates the goal that everyone in each round of the game should get the resource they need and thus survive to play again. The second is dangerous.

If some people give up and move to the side, wait to say "Switch!" until players make an effort to include outsiders. If the group doesn't make this effort, ask, "Is it acceptable to you that the success of some participants causes the failure of others?" Typically, that question will prompt some new efforts to help everyone find space.

If people start creating human pyramids, simple say that everyone must have their feet touching the ground, and everyone must be self-supporting.

At some point, people may ask whether their feet must be flat on the ground, or whether they can stand on tiptoe or position their feet in some other way and still follow the rules. A good reply is: "Everything that is not forbidden is permitted." Eventually, someone will realize that it is therefore legal for participants to sit or lie on the ground outside the circle with just, for example, their heels touching the ground inside the circle.

Notice how and with whom this idea originates and whether others in the group promote or resist it.

Generally, if a high-status participant, such as a senior manager or a tall male, initiates the idea, other group members will support it. The same idea coming initially from a low-status participant will often be ignored.

Step 7: When all group members have successfully managed to have their feet touching the ground inside the last remaining rope circle, the exercise is complete.

Say something like, "Terrific! You figured out how to give everyone space for living." Then lead a brief round of group applause. Help up those who are on the floor, and segue into the debrief.

Cautions

Normally, the close physical proximity that this game requires is not an issue for Western participants. However, if you believe that one or two people may feel uncomfortable with it, enlist them to pick up the ropes and to check for compliance with the rules. If you think more than just a few participants will be uncomfortable, do not use this exercise. You may possibly finesse this issue, when working with a non-Western group, by dividing the participants into a male and a female subgroup, so that there is no close proximity among those of the opposite sex.

Watch to ensure that no one creates a solution to the crowding that puts anyone under physical stress or poses the potential for someone to lose their balance and fall over. We have run **Space for Living** hundreds of times with no problem, but it is always wise to be cautious.

During the discussions and the debrief, it is best not to call on a specific person. Let the participants choose to share their thoughts, or not.

Variation 2

For this variation, provide enough loops for only about 80% of the participants.

Step 1: Create a straight line on the floor by using a length of rope. Ask participants to stand on one side of the line. Arrange the loops on the other side of the line, about 5'–10' away from the line.

Step 2: Ask participants to count off— "1," and "2," and "3," etc.—until each person has a number.

Step 3: Announce the rules of the game specified above for Variation 1. Then explain these additional rules:

- "With each round, only 'active' members will participate. I will indicate at the beginning of the round who is active."

- "After all the active members have found their space for living, I will check to make sure that they have followed the rules. I'll designate a new active group. Then I'll say 'Switch' again."

Step 4: Designate participants 1 and 2 active. Say, "Switch!" and let them find their space. Of course, they do this without trouble. Make a big show of checking that they've complied with the rules. Now announce that participants 1 through 4 will be active in the next round. That means numbers 1 and 2 remain in the field with the loops, and numbers 3 and 4 will enter the field when you say, "Switch!" Say it and wait for all four to find their place. On the next round, make participants 1 through 8 active. Continue in this way, doubling the number of active players each time until eventually all participants are active. When you get to that round, people will be forced to begin pairing up to accomplish the goal.

Step 5: Now continue as in Variation 1; that is, start removing a few loops with every round. Continue to designate everyone as active in each round and continue until you have reduced the number of ropes. Finally only the last and largest loop remains, and the group is forced to innovate in finding a solution.

Variation 3

Follow the same rules as for Variation 1, but say that anyone who has not found their own space for living after a few minutes will be asked to move to the sideline of the game. If people start being excluded by their colleagues, do not resist this. Just politely ask the excluded players to move to the sidelines and become observers. Reassure them that their observations will be useful at the end of the game, to help them stay engaged in the play.

Debrief

Depending on your group's degree of sophistication and goals, you can ask a variety of questions to prompt insightful discussions. However, as always, first give participants time to express their own views, feelings, and conclusions about the game. You can do this by asking, "Who has some view or feeling about the game that they would like to share?"

Then move to this set of related questions about underlying assumptions, paradigm change, control, and the ethics of inclusion:

- Did you assume at the beginning of the game that each person had to have his or her own loop? If so, why?

- Did you take the time during the game to discuss longer-term strategy? If not, why not? The typical response is that players felt the facilitator was pushing them from one round to the next by announcing "Switch." If you hear this reply, ask the next question.

- Who controlled the progress of the game from one round to the next? Participants generally assume that the facilitator has this authority. But you can point out that players had the final power. By simply standing on the rope, any one of them could have stopped the progress of the game long enough to hold any useful conversation within the group.

- How did those of you inside the circle feel about those who were not able to find a space? Who was responsible for outsiders' failure to find space?

- How did those of you outside the circle feel about those inside? Who was responsible for the fact that no one offered to help you find a place? If you play Variation 3, also ask those who were removed from the game how they felt.

- To succeed in this game, you need to experience two paradigm changes, or strategy shifts. First, you have to recognize that each person does not need to have his or her own loop. Second, you must figure out that your entire foot does not need to be touching the ground. In this exercise, how did these shifts occur? Who first had these ideas? Was it someone who already had a space? Someone who was excluded? What other characteristics did the initiators of these ideas have? How did other group members respond to these shifts? Did they support them? Resist them? If the group supported these changes, what was it about the initiators that caused others to support their ideas? Can you generalize from these conclusions to predict the most likely sources of new ideas in your organization?

- It was obvious early on that there would not be enough loops for everyone to find his or her own space using the policies with which you started the game. When the limits became obvious, did group members change immediately or wait to innovate until they had no alternative? If they waited, why? What are the costs of dealing with limits only after they are pressing hard on the system? How could you change the system to make it anticipate limits and innovate in advance of absolute necessity?

- The group processes you experienced in this exercise are quite common. Where have you seen them in real systems? In your own organization? What might you learn from this exercise that you could use in your organization to facilitate the spread of constructive new ideas?

Group Juggle

> "You must never tell
> a thing. You must
> illustrate it. We learn
> through the eye and
> not the noggin."
> Will Rogers

In our view, Will Rogers had it right: one of the best ways for a team to gain insight into the systemic nature of the challenges facing its organization is to illustrate those dynamic behaviors through first-hand experience. Since the 1960s, systems thinking workshop facilitators have used the Beer Game to provide groups with a hands-on experience of the notion that a system's structure drives its behavior[1]. In the Beer Game, a simple set of rules, combined with the participants' own desire to succeed, creates a cause-effect structure that generates unexpected behaviors. Delays and communication barriers within that system cause

[1] A description of the Beer Game and purchase information are available at: http://www.albany.edu/cpr/sds/Beer.htm

participants to magnify a small external signal—an unusually large order for beer—into a wildly oscillating pattern for inventories.

The Beer Game generates powerful lessons, but it takes at least an hour to play, and larger groups require numerous supervisors to ensure that players observe the right sequence of mechanical steps. **Group Juggle** takes about 10 minutes to play, and one facilitator can lead a large group. Moreover, the three-loop structure underlying the **Group Juggle** dynamics has many interesting applications in the realms of personal relationships and service enterprises.

In its initial set up, this exercise is similar to **Warped Juggle**. However, the instructions and debrief are substantially different. **Warped Juggle** explores the "Limits to Success" systems archetype, while **Group Juggle** focuses on the structures underlying exponential growth, overshoot, collapse and shifting dominance.

⇨ To illustrate the way a simple causal structure can produce complex behavior

⇨ To give participants the experience of being part of a system in which the identity of the dominant loop rapidly shifts

⇨ To reveal systemic structures that participants can identify with personally

⇨ To break down the formal, social barriers that exist in a workshop when its members first assemble

☆ Awareness that different groups can behave in similar ways when they are immersed in the same systemic structure

☆ Understanding of the structural relationships governing exponential growth, overshoot, decline, and shifting dominance

☆ First-hand experience with a systems archetype that explains overload and overshoot

Understanding of behavior over time graphs and causal loop diagram conventions

 Before members of a team can grasp the systemic causes of problems in their own organization, they need to feel confident that the system's behavior actually does result from factors that they can diagram and discuss as causal links and loops. In our view, some of the best ways to come to that understanding are through direct participation, and by reflection on personal experience.

For this purpose, **Group Juggle** is a valuable tool. It's also fun. Using a few balls or other tossable objects, it propels people up and out of their chairs and gets their blood circulating. This exercise can generate some real laughs; and it almost never fails to reveal profound new insights.

Caution

If a group member is unable to stand, you may try to conduct this exercise with all participants seated in chairs. If a participant is unable to catch or toss a ball or other soft object, you can ask that person to play the role of a "process observer," i.e., someone who provides invaluable commentary and feedback to the group during the exercise debrief.

To Run This Exercise

 Ideally, 15–20; if you have more than 20 players, split the group into smaller teams. The teams can do the exercise simultaneously, if you have enough supervisors. If not, lead each group through the exercise in sequence.

 15–60 minutes, depending on the number of lessons you want to convey, the level of participants' sophistication, and the length of the debrief

 Enough open space to let team members stand 3′ to 5′ apart in a circle. Objects will be tossed into the air, so the space needs a ceiling height of not less than 8′.

Equipment An overhead projector, flip chart, or white board; one ball or other tossable object per participant (e.g., a tennis ball or softball); and a box, shopping bag, waste basket, or other container to hold the balls

 Set-up Put a chair to your left or right. Put the balls in the container, then put the container on the chair, so it is easy to reach the balls without bending down.

Instructions

Step 1: Arrange group into teams. If your group consists of more than 20 people, divide them into smaller teams so no team has more than 20 members. Designate the other team(s) to serve as observers and select one team to begin. Stand with the beginning team in a circle. Ensure that members of the observing team stand far enough away from the circle that they won't impede people who will leave the circle to retrieve dropped balls. However, observers should stand close enough to watch the exercise unfold.

Step 2: Establish the throwing order:

- The goal during this part of the exercise is accuracy, not speed. Underhand throws are easier to catch. If anyone drops the ball, ask him or her to retrieve it and resume the sequence of throwing.

- Encourage people not to throw the ball to the person standing next to them. Instead, they should try to toss it to someone across the circle who still has his/her hands raised.

- Ask people to remember the identity of their catcher. This person will always be their designated catcher during the game.

- During the throwing process, ensure that no one gets the ball thrown to them more than once.

- Everyone holds their hands out in front of them at waist level, with elbows bent. You, the facilitator, throw the ball to someone. After that person catches the ball, he or she throws it to someone else, and then lowers his or her own hands. The person who just caught the ball throws it to someone else, and then lowers his or her hands too. Before each person throws the ball to someone else, they look for receivers whose hands are still at waist level. Continue until the ball has been thrown once to everyone, and everyone's hands are lowered. The last person in the initial sequence, when everyone else has now dropped their hands, should throw to the person who initially received the ball from you.

When that person has received the ball for the second time (the first time being from you), stop the action. The person who initially caught the ball from you will be the designated catcher of the last person in the team to get the ball. Thus, once a ball is thrown by you into the circle, it should continue to circulate around the group indefinitely, unless it is dropped.

- Get the ball that was in circulation and put it in the container.

Step 3: *Explain the rules of the game:*

- "Your team's goal is to keep as many balls in the air at the same time as possible by continually catching balls from your designated thrower and then throwing them to your designated catcher."

- "I can throw to anyone I see who is not currently holding a ball."

- "We'll start the game slowly. But as I see you successfully keeping more and more balls in the air, I'll throw more and more balls in." Make certain that everyone has heard this statement; it forms the basis for the driving causal loop in the early phase of the game.

- If there are observers, ask them to gather data about the number of balls in the air over the course of the exercise.

Step 4: *Test that everyone remembers their designated*

catcher. Ask team members to simulate throwing the ball by pointing in sequence to their designated catcher. You start the process by pointing at the person who received the ball from you. They, in turn, point to the person who is their designated catcher. Continue in this way around the circle. If anyone has forgotten the identify of their catcher, have the group figure it out. In rare cases, you may need to throw the ball around again to clarify the sequence or establish a new one.

Step 5: *Carry out the game.* Throw a ball to the first person you threw to during the sequence-determination round. As team members start passing the ball around according to the established sequence, wait 5 seconds. Then throw in another ball. Wait 3 seconds. Then throw in more and more balls. When people start dropping balls, usually after there are 10–15 balls in play, loudly urge the players to retrieve them. To provide even more distraction and give the group a few laughs, you can throw in a rubber chicken or some other bizarre but harmless object . As the chaos worsens, start throwing balls in rapidly to many different people, even if they're clearly not ready to catch them. Then call out: "Okay, stop!"

Step 6: *Switch teams.* If you divided the group into two teams, go through these steps again with the second team, asking the first team to observe the number of balls in the air. If there is only a single team, move to the flip chart or projector for the debrief.

Debrief

This exercise is rich in content. Below we sketch out the full recommended debriefing process. Guide your group through this sequence in a leisurely fashion, show them the relevant illustrations, provided below, and give players plenty of opportunities for questions and discussion.

Levels of Perspective

Start by explaining that it's useful to analyze a system at different levels of perspective.

We can think of these levels as including events, patterns of behaviors, structures, and mental models.

If we think of an iceberg, only ten percent of the ice—the event level—shows above the water line. Ninety percent of the iceberg remains hidden beneath the surface. Inquiring into the behavior of complex systems requires us to ask different questions at each of these different levels:

Now, ask participants to look at the **Group Juggle** exercise from different levels of perspective, for example:

Events

Ask questions that give participants the opportunity to share their general impressions of the activity:

- "What happened?"

- "What did you see happening during the juggle game?"

- "How did you feel?"

Typical responses:

- "There were different kinds of balls, and that made it tricky. Some weren't even balls at all."

- "I laughed when I saw the rubber chicken." "The balls sometimes collided. That made it impossible to catch them."

- "Some members of our team were really skilled at catching the ball; others were not."

- "Initially I could look around and watch the whole team. Later I just had to focus on my thrower and ignore everyone else."

Behavior over Time: Three Modes

Now segue into a discussion about behavior over time. Explain: "To take a deeper look at what happened, let's look beyond events to behavior patterns. At first, I threw in just a few balls, and you did a perfect job of juggling them. Then I threw in a few more, and still you did a great job. We can represent that behavior on a behavior-over-time graph."

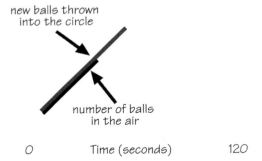

Ask:
"What was the pattern of balls in the air?" Give participants time to draw out the rest of the diagram on their own sheets of paper.

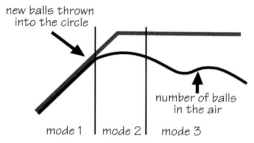

new balls thrown
into the circle

number of balls
in the air

mode 1 | mode 2 | mode 3

Explain:
"Once you began to drop the balls, I quit throwing any more into the circle. Notice that the pattern of behavior over time went through three modes:

- First was competence. As I threw in more balls, you caught more.

- I threw even more, and you caught even more. The second mode was limits to growth. I threw in more balls, but you did not catch more. What happened to the extra balls? They were on the floor being retrieved.

- The third mode—decline—came soon after that. Even though I was not throwing in more balls, your capacity to keep them in the air went down."

Ask:
"Why did this happen, and what could you have done to perform better?" Give players time to discuss and reply.

Typical responses:

- "At first it was easy, since there was lots of time to get ready for the next ball. Then they started coming faster, and I started to drop them."

- "I think if we practiced catching the ball, we could do better."

- "I think if all the balls had been the same, we would not have dropped so many."

- "Maybe we could have beat a drum or created in some other way a rhythm. Then we could all have thrown our balls at the same time."

Systemic Structures

Explain:
"Your ideas are useful, but it's difficult to know how they would work in practice. To figure that out, we have to move to the next level of perspective—systemic structure."

"Walk" participants loop by loop through the causal loop diagram shown below, starting with loop 1. The first loop is responsible for mode 1, exponential growth. The second loop produces mode 2, limits to growth. The third produces mode 3, the decline in participants' capacity to keep the balls in the air.

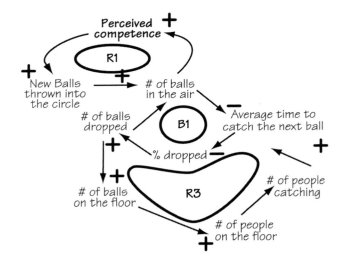

Explain:
"The behavior in this system changed from mode 1 to mode 2 to mode 3, because the dominance in the system shifted from loop to loop. Initially, loop 1 was

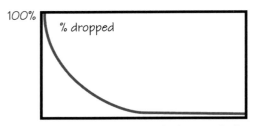

dominant, because your average time to catch the balls could decline without producing any change in the percentage of balls you dropped. That is, as I threw more and more balls into the group, you all caught them faster and faster without dropping any."

"But eventually the average time to catch the ball declined to the point where further reductions in that time, caused by my throwing in a few more balls, led to an increase in the percent of balls you dropped. Then loop 2 became dominant. But as the balls started to pile up on the floor, more and more of you left your places to retrieve them. That made loop 3 dominant, and the behavior of the system changed once again— this time, to a decline in your performance."

Ask:
"What could you have done to improve your performance?" Usually, people will suggest several policies to change the shape of the above graph, such as "Train people to catch better" or " Make all the balls standard, so they all feel the same" or "Count, so that everyone throws their ball at the same time" or "Some of us could throw high and some low to avoid collisions."

Explain:
"These are good ideas, but they wouldn't change the structure of the game; they would just shift the above graph down a bit. In other words, people might drop fewer balls for any given average time between catches, but the overall progression of modes wouldn't change." The changes would leave all the causal links in place. Those loops that are now reinforcing, remain reinforcing. Those loops that are balancing, remain balancing. Thus the proposals do not introduce any structural change. They just change the curves slightly; we would call them coefficient changes, because they change the numbers in the equation linking one variable to another.

Ask:
"What structural changes can you think of?" Give people time to respond.

Typical responses:

- "The facilitator could quit throwing in so many balls, when we start to drop them."

- "We could designate one person to catch all the dropped balls."

- "We could ask one of our members to stand outside the circle and understand the causes of our mistakes."

Explain:
"We could make two types of structural changes:

1. "Somehow eliminate the reinforcing loop that now controls the number of balls thrown into the circle. For example, I could suspend my assumptions about your ability to catch (Perceived Competence), which is based on the number of balls you are keeping in the air, and just stop throwing in new balls as soon as anyone drops a ball. This would break the link from Perceived Competence to Number of Balls Thrown In and thereby eliminate loop 1. It would also create a new balancing loop: As you drop more balls, throw fewer in. Fewer thrown in reduces the number of balls in the air. As number in the air goes down, the average time to catch the

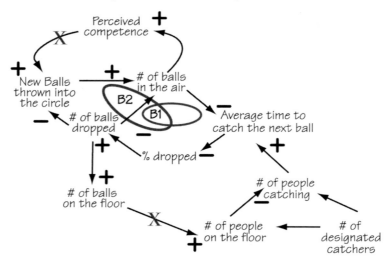

next ball goes up. That reduces the percentage, therefore the number, of balls dropped. As the number of balls dropped goes down, the number of balls on the floor also declines.

2. "Designate one or two people to do nothing but retrieve dropped balls, and tell all others in the circle to stay in place catching balls. With that new policy, an increase in the number of balls on the floor would no longer increase the number of people leaving the circle. Thus the number of people available to catch balls would not change. Breaking that link would eliminate loop 3 and therefore mode 3."

Mental Models

Ask:
"What mental models can you think of that may have influenced your performance" Again, give people time to respond.

Typical responses:

- "I thought we were trying to move the balls around the group as fast as we could."

- "I assumed we had to have all of our group members in the circle."

Systems Lessons

Group Juggle offers five useful lessons about systems:

- One can view a system from different points of view—data, behavior, structure.

- There is generally a shared and valid collective understanding about a system's mode and structure.

- The juggle behavior shows three different modes of behavior; these can be understood to result from shifting dominance in the system.

- An action that produces favorable results during

one mode of system behavior (i.e., when one loop is dominant) may be counterproductive during another mode of system behavior. For example, throwing in more balls during phase #1 produces more balls in the air. Throwing in more balls during phase #3 produces fewer balls in the air.

- Behavior comes from structure, not from coefficients. To get different behavior, you need to change the structure, i.e. add a loop or delete one.

Team-Learning Lessons

This exercise also provides five useful lessons about team learning:

- Think about a task ahead of time and identify the different roles that will be required. Assign them to the best qualified.

- Don't get caught up in the short-term events so much that you lose sight of the longer-term goal.

- Practice the task first, whenever possible.

- Take control of time.

- Decide on your capacity and keep the demands on your group below that, so you can maintain your quality standards. You can't handle an infinite work load. Something will limit the amount of work you can do. Either pick the limits, or the system will. Generally, the system's limits will come at a level that produces low quality.

Applications to Other Arenas

Overload isn't something that happens just to ball catchers. This same structure, identical loops with different names, can create havoc in many other areas of life. You can lead your participants into an extended, useful discussion by suggesting that they imagine how those three loops—competence, limits, and collapse— also illustrate behavior in personal relationships, or growth of a service firm.

Exercise Link

If you want to explore the phenomena of exponential growth further, you may choose to do **Group Juggle** in conjunction with **Paper Fold** and/or **Postcard Stories**.

resources

For more on the importance of looking at different levels within a system of interest, see:

Senge, P., N. Cambron-McCabe, et al. (2000). *Schools that Learn: A Fifth Discipline Fieldbook for Educators, Parents, and Everyone Who Cares About Education.* New York, Doubleday., p. 80.

Thinking in Levels: A Dynamic Systems Approach to Making Sense of the World, by Uri Wilensky and Mitchel Resnick. (1999) Journal of Science Education and Technology. Vol. 8, No. 1, p. 3–18.

Co-Authors

As a systems educator, writer, and researcher, Linda is dedicated to helping people of all ages develop their own insights about the nature of the complex, dynamic world around them. Linda is also the author of *When a Butterfly Sneezes: A Guide for Helping Children Explore Interconnections in Our World Through Favorite Stories* and *Connected Wisdom: Living Stories about Living Systems*. She focuses much of her work on making the concepts of tools of complex systems theory accessibly to a wide variety of audiences. Linda holds a doctorate in education from Harvard University and is the delighted mother of three children. (For more information, see: www.lindaboothsweeney.net.)

Dennis retired from university life in 2004 after 35 years as a professor and academic institute director. As President of the Laboratory for Interactive Learning he now develops games, writes books, speaks, and conducts teacher-training workshops to help professionals in many nations enhance their powers of communication and teaching. He has written or co-authored ten books, which have been translated into over 30 languages. Dennis has a doctorate from MIT in Management and Systems Dynamics.

The Illustrators

Visual images and inspiration by Nancy Margulis and Michelle Boos-Stone.

resources

Many of the resource materials referenced in this volume are available through Pegasus Communications:
Email: query@pegasuscom.com
Web: www.pegasuscom.com

To order **The Systems Thinking Playbook,** contact
Chelsea Green Publishing Company
Post Office Box 428
White River Junction, VT 05001
(802) 295-6300
www.chelseagreen.com